ISLAMIC NARRATIVE AND AUTHORITY
IN SOUTHEAST ASIA

CONTEMPORARY ANTHROPOLOGY OF RELIGION

*A series published with the Society for the
Anthropology of Religion*

Robert Hefner, Series Editor
Boston University
Published by Palgrave Macmillan

Body / Meaning / Healing
By Thomas J. Csordas

*The Weight of the Past: Living with History in Mahajanga,
Madagascar*
By Michael Lambek

*After the Rescue: Jewish Identity and Community in
Contemporary Denmark*
By Andrew Buckser

Empowering the Past, Confronting the Future
By Andrew Strathern and Pamela J. Stewart

*Islam Obscured: The Rhetoric of Anthropological
Representation*
By Daniel Martin Varisco

*Islam, Memory, and Morality in Yemen: Ruling
Families in Transition*
By Gabrielle Vom Bruck

*A Peaceful Jihad: Negotiating Identity and Modernity
in Muslim Java*
By Ronald Lukens-Bull

The Road to Clarity: Seventh-Day Adventism in Madagascar
By Eva Keller

Yoruba in Diaspora: An African Church in London
By Hermione Harris

*Islamic Narrative and Authority in Southeast Asia:
From the 16th to the 21st Century*
By Thomas Gibson

Islamic Narrative and Authority in Southeast Asia

From the 16th to the 21st Century

Thomas Gibson

palgrave
macmillan

ISLAMIC NARRATIVE AND AUTHORITY IN SOUTHEAST ASIA

First published in 2007 by
PALGRAVE MACMILLAN™
175 Fifth Avenue, New York, N.Y. 10010 and
Houndmills, Basingstoke, Hampshire, England RG21 6XS
Companies and representatives throughout the world.

PALGRAVE MACMILLAN is the global academic imprint of the Palgrave Macmillan division of St. Martin's Press, LLC and of Palgrave Macmillan Ltd. Macmillan® is a registered trademark in the United States, United Kingdom and other countries. Palgrave is a registered trademark in the European Union and other countries.

ISBN 978-1-349-53842-3   ISBN 978-0-230-60508-4 (eBook)
DOI 10.1057/9780230605084

Library of Congress Cataloging-in-Publication Data

Gibson, Thomas, 1956–
    Islamic narrative and authority in Southeast Asia : from the 16th to the 21st century / by Thomas Gibson.
      p. cm.—(Contemporary anthropology of religion)
    Includes bibliographical references and index.

        1. Islam—Indonesia–Sulawesi Selatan—History. 2. Sulawesi Selatan (Indonesia)—Religion. 3. Sulawesi Selatan (Indonesia)—Social life and customs. 4. Islam and culture—Indonesia—Sulawesi Selatan. 5. Ethnology—Indonesia—Sulawesi Selatan. I. Title.

BP63.152S825 2007
297.09598—dc22   2007061159

A catalogue record for this book is available from the British Library.

Design by Newgen Imaging Systems (P) Ltd., Chennai, India.

First edition: June 2007

10 9 8 7 6 5 4 3 2 1

Transferred to Digital Printing in 2008

Contents

Contents

List of Maps and Figures

Maps

Figures

Acknowledgments

The local scholar to whom I owe the largest debt by far was my host and mentor, Haji Abdul Hakim Daeng Paca. Among the many others who volunteered their time to instruct me in the finer points of Islam in South Sulawesi were Hama Daeng La'ju and Palippui Daeng Puga, masters of the arcane sciences (*ilmu*); Sirajang Daeng Munira, Alimuddin Daeng Mappi, and Muhammad Yakub Daeng Jagong, *Imams* of Ara; Muhamad Idris Daeng Buru'ne, *Imam* of Bira; Abdul Hamid Daeng Maming, former head of the Department of Education and Culture for Bonto Bahari; Daeng Pasau and Haji Mustari, *Kepala Desa* of Ara; Daeng Sibaji Daeng Puga and Muhammad Nasir Daeng Puga, reciters of *Sinrili' Datu Museng*; and Muhammad Idris Radatung Daeng Sarika, schoolteacher, master musician and former Darul Islam militant. I also owe a deep debt of gratitude to Rusnani Babo and Drs. Aminuddin Bakry, my hosts in Ujung Pandang. Dr. Abu Hamid, Professor of Anthropology at Hasanuddin University, provided me access to his seminar at the University and to the Indonesian academic community more generally.

My understanding of the way Islam has interacted with Austronesian symbolic systems has benefited from discussions with many fellow students of the area, including Benedict Anderson, Lanfranco Blanchetti-Revelli, Maurice Bloch, John Bowen, David Bulbeck, Ian Caldwell, Michael Feener, Ken George, Gilbert Hamonic, Robert Hefner, Michael Laffan, Michael Lambek, Ronald Lukens-Bull, Jennifer Nourse, Michael Peletz, Christian Pelras, James Siegel, Heather Sutherland, and Mark Woodward.

My first two visits to South Sulawesi in 1988 and 1989 were financed by a grant from the Harry Frank Guggenheim Foundation. The preliminary analysis of my findings in 1989 was financed by a Visiting Fellowship in the Comparative Austronesian Project of the Department of Anthropology, Research School of Pacific Studies, the Australian National University. Historical research in the Netherlands was supported in 1994 by a Senior Scholar award from the Fulbright Commission for lecturing-research in the Research Centre Religion and Society, University of Amsterdam, Netherlands. Further research and analysis was supported by the Southeast Asia Program at Cornell University, during a semester I spent as a Visiting Associate Professor in 1997 and during a year I spent as a Visiting Fellow in 2000–2001.

A Note on Makassar Names

In South Sulawesi, most people have several names. Everyone has a proper name acquired at birth. Nobles later acquire a title name composed of an honorific like *Andi'*, *Daeng*, or *Karaeng* followed either by another given name or by the name of a territory. Particularly devout individuals often prefer to use the Arabic name they are given when they are circumcised instead of their Makassar or Bugis name. The situation for high-ranking nobles is even more complex. They receive a birth name, a noble name, a series of territorial titles, and, in the case of the rulers of large kingdoms and empires, the title of Sultan followed by the name of a prophet like Ismail or by the phrase "servant of [one of God's attributes]," and a posthumous nickname. Thus the ruler of Tallo' who converted to Islam was I Malinkaeng Daeng Mannyonri Karaeng Kanjilo Karaeng Segeri Karaeng Matoaya Sultan Abdullah Awwal al-Islam Tumenanga ri Agamana, or I Malinkaeng Daeng Mannyonri, Lord of Kanjilo, Lord of Segeri, The Senior Lord, the Sultan Who Serves God, the First in Islam, He Who Sleeps in the Religion. For the sake of brevity, I have tried to use just one of each individual's names throughout the book.

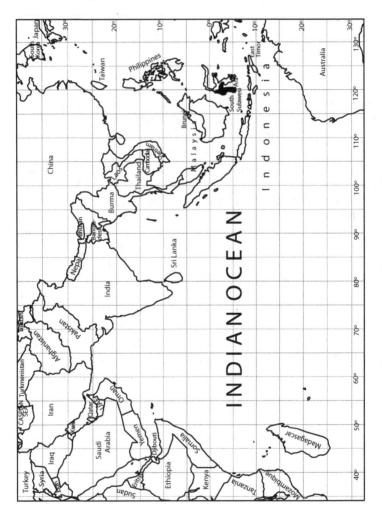

South Sulawesi in Relation to the Indian Ocean

Chapter 1

Introduction: Symbolic Knowledge and Authority in Complex Societies

This book is about the complex societies of Island Southeast Asia that converted to Islam between about 1300 and 1600 CE. For the most part, the members of these societies employ technologies, speak languages, perform rituals, and recount narratives that derive from a common Austronesian heritage. In a previous book, I explained how these shared forms of knowledge led to the development of a regional political economy in which a series of coastal kingdoms were loosely integrated through the long-distance exchange of material goods, royal spouses, and symbolic knowledge (Gibson 2005). In this book, I explore the way this regional system was transformed as it was integrated into a still wider system between 1300 and 1600 CE. Conversion to Islam played a key role in this process as it provided a cosmopolitan symbolic code that enabled Southeast Asians to marry, to trade, and to ally with fellow Muslims around the Indian Ocean. I also explore the implications of Island Southeast Asia's encounter with the predatory trading practices of Western Europe, an encounter that dates to the Portuguese conquest of Melaka in 1512.

By the end of the sixteenth century, the societies of Island Southeast Asia had been exposed to practical and symbolic forms of knowledge derived from every corner of the eastern hemisphere. To reduce this complexity to manageable proportions, I have drawn on the three ideal types of political authority formulated by Max Weber, and linked them to three ideal types of symbolic knowledge. These are Austronesian ritual knowledge and the traditional authority of hereditary kings; Islamic religious knowledge and the charismatic authority of cosmopolitan *shaikhs*; and documentary knowledge and the bureaucratic authority of modern states. Each of these spheres of knowledge and authority should not be thought of as relating to a stage in an evolutionary sequence of social types, but as relating to a distinct set of social and psychological processes that have coexisted and interacted with one another in Islamic Southeast Asia for several centuries.

The point about the epistemological and political complexity of these societies can be made quite concretely by reference to Ara, the Makassar village in South Sulawesi where I conducted fieldwork in the late 1980s. At that time, many of the older men were literate in three scripts and fluent in at least three languages. An Indic syllabary known as the *lontara* was used for recording noble genealogies and the chronicles of royal ancestors in Konjo, the local dialect of Makassar. Arabic script was used for reading and writing sacred Islamic texts in both Arabic and Konjo. Roman script was used in schools and by state functionaries for writing official documents in Bahasa Indonesia, the form of Malay that was used by the Dutch to administer their colonial state and that was adopted as the national language at the time of independence.

Among the Makassar, the traditional Austronesian political order was headed by a king who was descended from mythical otherworldly beings. A system of social rank that was determined by genealogical proximity to the founding royal ancestor divided the entire population into a finely graded hierarchy. Claims to social rank were made and contested during an elaborate series of life-cycle rituals marking birth, adulthood, marriage, and death. By participating in these rituals, individuals came to see themselves as occupying differentiated niches within a hierarchically organized social whole. It formed the basis of what Weber would have called the traditional authority of the Makassar kings.

A very different set of symbolic practices based on Islam has coexisted with this Austronesian system since the seventeenth century. All Muslims perform certain life-cycle rituals that detach people from their families of procreation and integrate them into the *umma*, a cosmopolitan community made up of ethical individuals whose ultimate worth can only be gauged by the Supreme Being on the Day of Judgment. A dedicated few engage in more demanding religious practices, which generate a very different kind of ranking system than the one generated by Austronesian rituals. These elite practices include the pilgrimage to Mecca; the recitation, memorization, and mastery of commentaries on sacred scriptures by religious scholars; and training in mystical disciplines by Sufi masters. Throughout the premodern era, the cosmopolitan community of Islam was knit together by the religious elites who engaged in these practices. Even today, lowborn men of talent and ambition can acquire a kind of charismatic authority during their travels that trumps the traditional authority of the nobility and the bureaucratic authority of the state.

Even before their conversion to Islam, the Makassar kings encountered the ruthless practices of Portuguese mercantilism. They encountered them in an intensified form with appearance of the Dutch East India Company (VOC) in the seventeenth century. The VOC was one of the first truly global corporations, dedicated to the rational maximization of shareholder value. The means by which this was achieved constituted a set of symbols, disciplines, and institutions no less elaborate than the ones associated with Austronesian kingship and Islamic sainthood. The VOC was structured as a bureaucratic hierarchy open to merit. A motley assortment of mercenaries and sailors who had been press-ganged in all the ports of Europe was turned into a reasonably efficient war machine through the application of rigorous military drill. Relevant economic and political information was collected, recorded, and archived by a well-trained corps of merchants and clerks. The significance of the personal qualities and social networks of high-ranking officers was minimized by regularly rotating them throughout the territories controlled by the VOC.

Although the Makassar empire of Gowa had developed a reason-ably efficient bureaucratic structure of its own by the beginning of the seventeenth century, its legitimacy continued to rest on the centrality of the ruler to the traditional system of royal rituals and to the cos-mopolitan networks of legal scholars and mystical masters that had come with the conversion of the king to Islam. The internal cohesion of the political elites of the empire continued to depend on a complex network of marital alliances among them. The impersonal officers of the VOC were unable to participate in any of these ritual, religious, or marital practices, and so their power lacked any kind of legitimate authority among their local subjects.

During the eighteenth and nineteenth centuries, Makassar rulers and subjects alike tended to react to the presence of this novel form of power in their midst with incomprehension and hostility. As a result, South Sulawesi was in an almost constant state of rebellion throughout this period. It was not until the Dutch began to train Makassar people to serve as officials in the late colonial state that documentary knowledge and bureaucratic authority were absorbed into the local social formation. Between about 1910 and 1950, Islamic and bureaucratic forms of knowledge interacted to produce a revolutionary new model of political authority, a form of Islamic nationalism that demanded the replacement of both traditional nobles and foreign colonial officers by Indonesians learned in the *shariah* law. Between 1950 and 1965, Islamic militants fought to establish the Republic of Indonesia as an Islamic state, *Darul Islam*. During Suharto's New Order regime (1965–1998), politically motivated Islamic movements were suppressed. Islamic modernism took its place alongside the traditional social hierarchy and the authoritarian nation-state as just one set of symbolic practices among others.

My analysis of the interaction of Austronesian, Islamic, and bureaucratic knowledge and authority in Island Southeast Asia has theoretical implications for the study of complex societies in general. A similar level of complexity exists in any society whose members are embedded in a local kinship system, owe allegiance to a regional polity, and adhere to a "world religion." For example, the societies of Northwest Europe use languages and kinship systems that derive from the ancient Germans, political and legal systems that derive from the Roman Empire, and religious systems that derive from the prophetic traditions of the eastern Mediterranean. Just as in Island Southeast Asia, European social, political, and religious practices derive from different sources, are reproduced through different mechanisms, and generate different kinds of individual agency.

In order to analyze the interaction of traditional, prophetic, and bureaucratic forms of knowledge and authority, I have found it necessary to draw on several distinct theoretical traditions. My analysis of the implicit forms of symbolic knowledge associated with traditional authority derives from the work of Émile Durkheim and Claude Lévi-Strauss on the collective rituals, myths, and kinship systems of nonliterate societies (Durkheim [1915] 1995; Lévi-Strauss 1966). My analysis of religious forms of symbolic knowledge and of charismatic authority derives from the work of Friedrich Nietzsche and Max Weber on the explicit doctrines and ascetic disciplines inspired by the visions of individual prophets in literate societies. My analysis of documentary forms of knowledge and of bureaucratic authority derives from the work of Max Weber and Michel Foucault on formal institutions such as states, corporations, factories, and schools.

I have found that I must modify each of these theoretical traditions to correct for the fact that they were developed on the basis of a small sample of societies, and that they tend to assume that one form of knowledge and authority replaces another in an evolutionary sequence, not that they coexist and interact with one another through time. I will begin by stating in summary form the general theoretical approach I have developed on the basis of these authors. Since these authors are used in anthropology, sociology, comparative religion, and social history in many different ways, I go on to say something about how I have adapted them for my own purposes.

Symbolic knowledge draws on a wider pool of practical knowledge about the natural and social world that is largely prelinguistic and implicit. Symbolic knowledge is communicated through material symbols; embedded in human bodies through disciplinary practices; and transmitted and transformed within concrete social institutions. I call the set of symbols, disciplines, and institutions through which a specific kind of symbolic knowledge is reproduced a "symbolic complex." Different symbolic complexes tend to generate different internal experiences of the self and external experiences of society. Defined in this way, symbolic complexes cannot be explained as epiphenomena of an underlying level of social reality such as economic production or rational self-interest. Social formations are instead made up of a multiplicity of interacting symbolic complexes, each of which has a relative autonomy from the others and an independent genealogical origin.

I use the term "ideal model" for the way the experiences and relationships generated by the symbolic complexes present in a social formation are brought to consciousness and synthesized into an explicit model, which then becomes the goal of political action. The power of rulers within a social formation largely depends upon the legitimate authority that is granted to them by their subjects when they act in accordance with a widely accepted ideal model. When the ideal model generated by one symbolic complex no longer makes sense of a social formation undergoing rapid historical change, the legitimacy of the entire social formation may be called into question. This can result in a "revolutionary" situation in which social actors consciously seek to replace one ideal model of legitimate authority with another. More commonly, social formations undergo gradual transformation as competing groups of actors attempt to establish rival ideal models as dominant in social and political life.

The commitment of an individual to a particular ideal model may be more or less sincere. In the course of my analysis, I shall have occasion to describe many situations in which political actors appear to engage in the cynical manipulation both of explicit models and of implicit symbolic complexes in order to maximize their material wealth and power. I will reserve the term "ideological manipulation" for this sort of action. Defined in this way, ideological manipulation is always parasitic on the ideal models that actors have developed on the basis of their participation in the myriad symbolic complexes found in their society. It can only be effective when it appeals to a set of subjective experiences and dispositions that are beyond the conscious control both of manipulators and of their target audiences.

In the first few chapters of this book, I will be dealing primarily with conscious political models as they are articulated in a series of allegorical narratives that reconcile Islamically inspired ideals with the traditional social structure and with the bureaucratic state. As the material at my disposal becomes richer in later chapters, I am able to pay closer attention to the level of self-conscious manipulation of competing ideal models by opportunistic political actors. I deal with the largely implicit level of knowledge of the self and society generated in traditional, Islamic, and bureaucratic rituals in a separate book.

Durkheim, Lévi-Strauss, and Traditional Authority

I was originally trained at the London School of Economics in a tradition of social anthropology that had little use for the romantic German–American concept of "culture." It drew instead on the French tradition of symbolic analysis pioneered by Émile Durkheim, Claude Lévi-Strauss, Daniel Sperber, and Maurice Bloch. In North America, this tradition is usually labeled "structuralism." So many misconceptions have become attached to this label that I usually avoid it. In my view, the North American understanding of "structuralism" in fact refers to a variant of Durkheim's sociology of knowledge that was developed in Leiden during the 1920s and 1930s by Indonesianists such as J.P.B. de Josselin de Jong, W.H. Rassers, and F.A.E. van Wouden (P.E. de Josselin de Jong 1977, P.E. de Josselin de Jong and Vermeulen 1989). It attempted to correlate static patterns of symbolic classification with social groupings and was indeed ahistorical and devoid of human agency. This variant was taken up during 1950s by the students of E.E. Evans-Pritchard during his time at Oxford (Cunningham 2000: viii). For reasons I have discussed elsewhere, I would include among these students Rodney Needham, Louis Dumont, and Mary Douglas (Gibson 1989; for representative examples of "static structuralism," see Needham 1958; Dumont 1972, 1975; Douglas 1973). Needham was one of the original translators of both Leiden structuralism and of Lévi-Strauss into English (Lévi-Strauss 1964; van Wouden [1935] 1968). This had the unfortunate result that Needham's voluminous attempts to correlate eastern Indonesian kinship terminologies with archaic marriage systems were taken by many as representative of Lévi-Strauss's form of "structural analysis," when in fact they were closer to that of the Leiden school.

In my view, a careful reading of Lévi-Strauss's entire corpus demonstrates that the search for static patterns is of no interest to him whatsoever. He makes it very clear that he views social formations as made up of highly heterogeneous assortments of symbolic systems that behave very differently in time, depending on whether their material substrates are human beings (kinship and totemism); goods (economies); or language and symbolic knowledge about the world (mythology). He is interested not in the surface patterns that may be observable in these subsystems at any one point in time, but in the way the real conflicts and contradictions between them generate an unending series of attempts to reconcile them in thought through a relatively small number of cognitive processes. It is because of his interest in these processes that many of his British and French followers such as Bloch and Sperber eventually became more interested in cognitive psychology than in ethnography as an end in itself (Sperber 1985; Bloch 1998).

There is thus nothing static about Lévi-Strauss's version of structural analysis. Indeed, structural analysis can only be applied to the transformations that occur between one symbolic subsystem and another, and to the symbolic system as a whole as it evolves through time. In some of his later works, Lévi-Strauss sketched out an ambitious theory of historical transformation (1982, 1983, 1987). He distinguished between kinship-based societies, where accumulations of wealth and power are continually redistributed throughout the whole through institutions of marriage, gift exchange, and other forms of reciprocity; house-based societies, in which these institutions become ideological covers for the accumulation of wealth and power; and class-based societies, which are openly dedicated to accumulation. It is this dynamic and historical aspect of Lévi-Strauss's work that I find most relevant to my own concerns.

In a series of studies, Marshall Sahlins has used a dynamic version of Lévi-Strauss's structuralism to analyze the encounter of the societies of eastern Austronesia with European colonialism (Sahlins 1981, 1985, 1995). The relatively isolated kingdoms of the Pacific discussed by Sahlins provide us with some of the only documented cases of pristine states. They are a poor model for the theorization of social change in other parts of the world precisely because they were so isolated from one another and from societies belonging to different ethno-linguistic traditions. By contrast, the western branch of the Austronesian diaspora, the islands lying along the sea lanes from East Africa to East Asia, provides an example of cultures that have been interacting with virtually every cultural tradition in the eastern hemisphere for centuries. In my previous book on Austronesian kingship in Southeast Asia, I explained in detail how and why Lévi-Strauss's approach to the interpretation of Austronesian kinship systems, rituals, and myths illuminates the history of the formation of a regional political system in western Austronesia (Gibson 2005). Here I can only provide a brief outline of that analysis.

Austronesian rituals, myths, and royal chronicles relating to noble life-cycle rituals and to the origins of maritime trade, agriculture, warfare, and kingship were used by royal houses all around the Java Sea to legitimate their traditional authority between 600 and 1600 CE. Traditional rituals draw on the whole range of practical or encyclopedic knowledge of the natural and social environment that people build up in the course of everyday life. They operate on all five senses simultaneously to convey a complex set of implicit meanings. In South Sulawesi they usually include elaborate visual displays of pungent foods, collective chanting, stylized gestures, and traditional costumes that highlight sexual differences. Repeated exposure to the same ritual sequences establishes an increasingly intricate network of associations between different domains of experience. The multidimensional correlations that are established through these media transcend the linear sequences to which verbal commentaries are restricted. Such commentaries can thus never exhaust the potential meanings implicit in ritual performances (Sperber 1975). As a result, symbolic systems that are based primarily on ritual performances will only be fully intelligible to senior members of the small local communities that perform them.

In ritual contexts where the reproduction of the local social structure is paramount, the most salient characteristics of an individual are ascribed qualities such as age, gender, familial affiliation, and hereditary rank. Makassar social structure is conceptualized in terms of the relationship between "houses" that are ranked in a hierarchy based on their ties to hereditary political rulers. An individual's place in this system is defined in terms of his or her inheritance of social rank from both parents. Women are expected to marry partners of equal or higher rank. The geographical extent of the relevant social structure depends on the level of a social actor in the social hierarchy. It is relatively restricted for those at the bottom of the hierarchy, but it is potentially coextensive with the entire political region for those at the top. The traditional knowledge and authority that is the basis for the local social structure is transmitted primarily through the implicit ritual symbolism employed during life-cycle rituals. Such rituals can only be investigated through participant observation in the field, which I conducted in the village of Ara, South Sulawesi in 1988, 1989, and 2000.

Makassar social structure generates individuals who are motivated to claim a higher social rank for their family than the one to which it was previously entitled. To succeed, they must persuade other members of the local community to accept a revision of their family's genealogical narrative, for a person's rank derives from that of his or her ancestors. A common ploy for a man who has acquired great wealth is to offer a large sum of bridewealth to the family of a woman who has a higher rank. If they accept it, the bride's family must tacitly acknowledge the man's claim to a higher rank since by definition husbands must be of equal or higher rank than their wives. The whole community must then be persuaded to accept the new state of affairs by participating in an exceptionally elaborate sequence of wedding rituals. Strategies in this arena often play themselves over the course of an entire lifetime, or even through several generations in the case of the highest-ranking noble houses.

Cosmological myths have many of the same characteristics as life-cycle rituals. They have no authors but are passed on from one generation to the next. The symbolic oppositions at work within them operate largely at an unconscious level and are closely related to the ones at work in ritual. While they unfold on the surface as linear narratives, myths are actually more like tightly structured musical scores in which each element plays a predetermined part in a structured whole. Since myths are expressed in the relatively explicit medium of language they are easier to translate into the symbolic codes of neighboring communities and typically circulate within broader geographical regions than rituals. While myths undergo systematic transformations as they adapt to each local situation, they also provide a shared symbolic code through which neighboring communities can communicate with one another. This is especially important as local societies become organized into larger political units and as political units attempt to establish diplomatic and economic relationships with one another within larger regions.

Royal origin myths uphold traditional authority by relating historically contingent institutions such as kingship, tribute, trade, and warfare to relatively stable domains of human experience, such as the cosmic cycles of the sun, moon, and stars; the meteorological cycles of the seasons; and the biological cycles of plants, animals, and humans. These mythical correlations between the temporal order and the cosmic order are intuitively persuasive to both rulers and subjects because they all participate in rituals that establish similar correlations at the level of immediate sensory experience.

In ritual contexts where the reproduction of traditional power relationships was paramount, the identity of individuals as subjects of a particular ruler was their most salient characteristic. Subjects declared their loyalty by attending the installation of a new ruler, by periodically paying tribute to the ruler, and by swearing an oath of loyalty to a ruler before going into battle.

The royal houses of South Sulawesi have intermarried since the seventeenth century, and no understanding of Makassar social structure is complete without a detailed analysis of the extensive royal chronicles and genealogies that record the political and marital alliances of these rulers. When studying the highest levels of the social hierarchy, field-work must thus be supplemented by extensive comparative reading on the ethnography and history of the whole region. Over the past fifteen years, I have examined many such texts in Makassar and in Indonesian, Dutch, French, and English translations.

Rituals and myths map different domains of experience onto one another to form complex, multidimensional wholes that are extremely stable over time, since disruptions caused by changes in one domain of experience tend be brought back into line with more conservative parts of the symbolic whole. They relate to universal aspects of human experience, expressed through symbols drawn from particular local conditions. As Lévi-Strauss has shown, they return again and again to the relationship between nature and culture, between male and female, between incest and exogamy, between identity and difference. They draw on a fundamentally cyclical model of time in which every significant event, such as a noble wedding ritual, is interpreted as a more or less imperfect copy of an original event, such as the mythical wedding of the founding royal ancestor. They convert the universal human experience of durational time into the eternal return of bio-logical regeneration. They portray the individual as a component of larger social wholes such as local descent groups, noble houses, and kingdoms. In South Sulawesi, they identify an aspect of the self that survives the current life cycle in order to reappear as a component of the encompassing group in a later generation.

For Lévi-Strauss, myths are characteristic of "cold societies," egalitarian societies in which struggle between economic classes has not yet begun to generate endogenous social change. Cold societies change as a result of exogenous events generated at the level of the techno-environmental infrastructure, but they react to such dise-quilibrating forces by trying to return to the status quo ante. Since symbolic knowledge in these societies is passed on primarily through ritual performance and oral recitation, subtle adjustments in the system can continually be made without anyone being conscious of them.

By contrast, full-fledged historical narratives are characteristic of "hot societies," stratified societies in which one class endlessly accumulates wealth and power at the expense of another. In hot societies, the constant accumulation of wealth and power in the present leads to the continual transformation of social institutions. Because of the presence of writing, people become conscious of the ever-growing difference between the present and the past. In order to legitimate current institutions by reference to the circumstances of their founding, historical narratives must be continually revised. Thus in 1962 most contemporary French political institutions and debates continued to trace themselves back to the Revolution of 1789, while the issues at stake in the Fronde rebellion of the seventeenth century had lost their relevance to Frenchmen's lived history (Lévi-Strauss 1966: 254–256).

In between kin-based and class-based societies lie societies organized in terms of noble houses, social entities that continue to be loosely based on the idiom of kinship but which are actually engaged in a ruthless contest for wealth and power. These noble houses differ from the lineages in a closed system of marriage exchange that seek to reproduce themselves and the system to which they belong from one generation to the next. Noble houses have cumulative histories and seek to pass on whatever wealth, power, and prestige each generation manages to accumulate to whichever heir seems most likely to preserve it (Lévi-Strauss 1982; Gibson 1995, 2005; Joyce and Gillespie 2000). Among the Maksassar, dynastic genealogies and royal chronicles recorded all of the titles, lands, and subjects that noble houses managed to accumulate over several generations. Until their conversion to Islam in the seventeenth century, the Bugis and Makassar peoples of South Sulawesi would appear to have fallen squarely within Lévi-Strauss's definition of *sociétés à maison*.

Nietzsche, Weber, and Foucault on Charisma and Bureaucracy

The focus of this book is on the way Islamic concepts of knowledge and authority introduced from Southwest Asia have interacted with the Austronesian symbolic complexes analyzed in my previous book and with the bureaucratic structures introduced from Western Europe. Since authors such as Lévi-Strauss and Sahlins have little to say about the symbolic complexes that lie at the heart of prophetic religions and bureaucratic institutions, I supplement their approach to symbolic knowledge with one that derives from authors such as Nietzsche, Weber, and Foucault.

Again, my reading of this tradition may appear somewhat idiosyncratic since these three authors are often embraced or rejected on the basis of their preoccupation with the individual's will to power and search for meaning. In my opinion, they viewed the "sovereign subject" not as a universal quality of human beings, but as a historically contingent by-product of specific historical developments, such as the spread of Christianity in the Roman Empire (Nietzsche 1966), the Protestant Reformation and capitalist rationality (Weber 1985), and the Catholic Counter-Reformation (Foucault 1978, 1991: 88).

More generally, Weber held that while most human conduct is guided by custom and tradition, a significant part is guided by the making of self-conscious choices among competing goals. He took the existence of the ability to make such rational choices as what set human action apart from animal behavior, and as what required the development of a different set of methodologies in the human sciences from those employed in the natural sciences (Weber 1975). The "science of social action" consisted in the study of the social mechanisms that generated a proclivity on the part of social actors to make rational decisions by reference to explicit future goals. In his writings, these mechanisms boil down to two: those that promote actions oriented to ultimate ends, usually some form of religious salvation advocated by an ethical or exemplary prophet; and those that promote actions oriented to the efficient achievement of some easily measured worldly outcome, usually political power or economic profit. The pursuit of the former corresponds roughly to what I will be calling prophetic knowledge and charismatic authority, the latter to documentary knowledge and bureaucratic authority.

In his historical analyses, Weber tried to show how these two forms of knowledge and authority interacted with and reinforced one another. The idea that an individual should live his or her entire life in accordance with a consistent ethical code is likely to be most appealing to those whose everyday occupations reward hard work and consistent effort. Thus the messages of ethical and exemplary prophets tended to have their greatest appeal among urban artisans and merchants, whose success depended on their own efforts to a greater degree than did successful harvests on the efforts of peasants, or military victories on the efforts of warriors. As agriculture and warfare were mechanized and made as predictable as manufacturing and commerce, the sphere of human experience guided by unconscious, traditional meanings tended to contract while that guided by conscious, rational meanings tended to expand.

Too many social theorists who draw on the tradition of Nietzsche and Weber miss the point that making rational choices is not a universal feature of human conduct, but the product of very specific sets of symbols, disciplines, and institutions (see Gibson, 1989 on Ortner 1984). If anything, Nietzsche, Weber, and Foucault were overly preoccupied with the relatively recent appearance and imminent dissolution of the "sovereign subject," whether at the hands of mass culture, the iron cage of capitalist rationality, or the spread of disciplinary forms of power. They were each guilty in their own way of assuming that new symbolic complexes would necessarily extinguish old ones, that the relational selves generated by kinship and the ethical selves generated by religion would be replaced by the impersonal, anonymous selves generated by bureaucracy.

Where I find Foucault's work an advance on that of Weber is in his insistence that different forms of subjectivity are not just spontaneously produced by the ordinary experiences of everyday life in an occupation. They are created and nurtured by the "power-knowledge" relations that are embedded in the specialized symbolic systems, disciplines, and social institutions that I have called symbolic complexes. Foucault views power and knowledge as being tightly linked, so that "the subject who knows, the objects to be known and the modalities of knowledge must be regarded as so many effects" of "power-knowledge" (Foucault 1977: 28). In my view, this way of formulating the relationship between knowledge and power is much too direct, even teleological. I would say that while all forms of knowledge have implications for the exercise of power, these implications are usually complex and contradictory. For example, colonial schools did not just produce a new kind of human object that could be manipulated at will by the ruling class. They also produced a new kind of human subject that was capable of organizing resistance against foreign rule far more effectively than the charismatic warriors produced by premodern institutions. In the Islamic world, the graduates of modern schools and *madrasas* have been just as likely to rebel against their masters as to acknowledge their authority (Mitchell 1988; Starrett 1998; chapter 7 of this book).

Foucault also tended to underestimate the extent to which premodern social formations relied on elaborate disciplinary techniques to produce specialized social agents such as religious mystics, skilled warriors, and master craftsmen. In his discussion of premodern social formations in *Discipline and Punish*, Foucault confined himself to a simple contrast between the practical disciplines of the modern age and the purely symbolic displays of royal power characteristic of the *ancien régime* (Foucault 1977: 130). Our understanding of the role ascetic rituals played in the creation of particular kinds of disciplined subjects in premodern societies is still in its infancy (see Mauss [1934] 1973 for an early statement of such a program). Talal Asad has gone some way toward analyzing its role in medieval Christianity (Asad 1993: 62). There is also a growing body of work on the disciplinary effects of religious ritual in Islam (Messick 1993; Bowen 1987, 1993; Lambek 1993; Gade 2004). On the basis of this work, it is clear that bodily disciplines were no less important in premodern societies than they are in modern societies.

As I noted earlier, I will take up the role of traditional, Islamic, and nationalistic disciplines in producing Austronesian, Islamic, and modern forms of subjectivity in a future book. Here I will simply suggest that all the great cosmopolitan religious traditions endeavor to construct sovereign, ethically autonomous selves in opposition to the relationally defined persons constructed by traditional rituals and to the "rationalized" actors constructed by bureaucratic disciplines. Different forms of human subjectivity are just as socially constructed and just as geographically and temporally variable as the social institutions in which they are embedded. Many specialized forms of subjectivity are produced in a single society through diverse disciplinary practices such as kinship ceremonies, religious rituals, military training, and formal schooling. In "modern" societies, a bureaucratic sense of the rational individual exists in a state of tension with a charismatic sense of the ethical soul and with a traditional sense of the socially situated person. The complexity of modern societies is reflected in the complexity of the modern self. But this complexity extends back in time at least to the "axial age" (800–200 BCE), when the great prophetic religions were founded as counterpoints both to the obligations owed to local kinship groups and to the imperial bureaucracies of the day (Hodgson 1974 I: 112).

Another major contribution of the Weberian tradition lies in the way it reformulated the question of power in terms of that of legitimate authority. Classical Marxism dealt with the problem of authority by assuming that worldviews were spontaneously generated by the class-specific conditions of everyday life. During periods of rapid social change, these conditions tended to produce two great competing worldviews based on the predominant property relationships at that moment in history. In response to the growing reformism of Western European workers and to the radicalization of Russian workers, Vladimir Lenin introduced the idea that the consciousness of the working class could only be raised through the efforts of professional revolutionaries trained in the scientific understanding of capitalism (Lenin 1978).

In response to the rise of Fascism in Europe during the 1930s, Antonio Gramsci introduced the concept of hegemony to explain the similar failure of the Italian working class to spontaneously produce a worldview counter to that of the ruling classes. He explained this failure by stressing the importance of extra-economic institutions such as the state, church, family, and school in producing ideologies that normalized the existing distribution of power. The creation of a counter-hegemonic ideology would inevitably be the result of a protracted series of dispersed struggles within each of these institutions, a process he called a "war of position" (Gramsci 1971).

Weber's notion of authority was similar to Gramsci's notion of hegemony. Both were concerned with what makes the exercise of power by the few over the many both meaningful and acceptable to the latter. For Weber, what is interesting is not some universal drive to dominate and to resist, but the way particular political arrangements are justified by reference to larger frameworks of meaning. It is only illegitimate power that is resisted, and all forms of power are far from illegitimate. Further, a form of power that loses its legitimacy can only be resisted in the name of some alternative form. Cosmopolitan religious traditions often provide an alternative symbolic language in which resistance to a regional political order can be articulated. It is here that Weber's pluralistic view of social causality is able to transcend even Gramsci's relatively sophisticated discussion of hegemony. Weber was free to contemplate the interaction of any number of symbolic systems, disciplines, and social institutions without reducing one to another. As a good Marxist-Leninist, Gramsci tried to relate everything that happened on the ideological battlefield back to the fundamental contradictions generated by the economic infrastructure.

On the question of authority and consent, Foucault often reverts to a concern with an abstract form of power and resistance that is closer to Nietzsche than to Weber or Gramsci. He seldom seems to contemplate the possibility that legitimate power or authority might not only exist, but that its existence might be the normal state of affairs. In this, Foucault has much in common with contemporary authors such as James Scott who deny the possibility that subordinates might ever accept the legitimate authority or hegemony of their rulers (Scott 1985). They assume the existence of autonomous individuals who exist independently of any particular set of symbolic complexes and the specific forms of subjectivity they generate. This assumption seems to me a real theoretical retreat in relation to the work of Weber and Gramsci.

In summary, socially and historically variable forms of subjectivity, knowledge, and authority are generated by a concrete array of contradictory symbols, disciplines, and institutions. Complex social formations are irreducible to the operation of a single underlying mechanism. They are composed instead of a range of symbolic complexes each of which interacts with the others according to its own internal logic. Resistance and consent are the products of a complex interaction of all the symbolic complexes present in a particular time and place.

Islamic Narrative and Charismatic Authority

What we now think of as the "world religions" originated during the "axial age" with the appearance of multiethnic cities inhabited by significant numbers of literate merchants and artisans. Under these conditions, the implicit meanings of traditional symbolic systems became objects of conscious reflection. They were reformulated and rationalized by individual prophets who imposed a systematic personal meaning on them. These meanings were recorded in scriptures that served as a template for later followers, who strove to achieve the same degree of personal autonomy and ethical integrity as their founding prophet. This process involves a protracted, arduous practice of mental and physical self-discipline, structured by the symbolic system inscribed in the scriptures. These disciplines often required the deliberate abrogation of the sense of relational personhood inculcated by the rituals of kinship and marriage. Hence the tendency for prophetic religions to regard celibacy, mendicancy, and austerity as helpful, if not necessary, in achieving salvation.

Virtually all of the Austronesian peoples of Island Southeast Asia became integrated into the Islamic world between 1300 and 1600 CE. This usually occurred when cosmopolitan Islamic experts persuaded a local ruler that privileged royal access to Islamic knowledge would provide them with a source of charismatic authority they could use to govern their ever-expanding territories and to interact as equals with Islamic rulers around the Indian Ocean. (As I explain in chapter 2, I use the concept of "charismatic authority" in a much narrower sense than Weber did.) Traditional rituals and myths that drew their power from unchanging features of a local environment were limited in their power to legitimate large-scale political units.

During the axial age, large-scale tributary states that aspired to become world empires often turned to sacred narratives with pretensions to universal validity to consolidate their legitimacy. These sacred narratives used the extraordinary experiences of charismatic individuals to construct an ideal model of human conduct that was not tied to any one time or place. They did so by interpreting the unique life experiences of the prophet as the manifestation in the temporal world of a transcendental order of being, thus bridging the gap between human individuals and the cosmos (for general discussions of the role of narratives in social life, see Niles 1999; Ochs and Capps 2001).

The religiously inflected narratives I discuss in this book are situated somewhere between the royal chronicles of regional polities and the national histories of fully formed class societies. Unlike the divine beings at the center of royal origin myths and the hereditary rulers at the heart of royal chronicles, the protagonists in these religious narratives are typically humble individuals of obscure origins who reveal their charismatic qualities through their response to extraordinary events. Their deeds often end in the transformation of the existing symbolic system, either internally, as when an individual experiences a spiritual epiphany or conversion, or externally when they succeed in inaugurating a social order based on a new set of institutions. The archetypical protagonist in this kind of narrative is the ethical prophet who is called upon by a higher power to denounce the existing state of affairs as out of line with the transcendental order.

For Makassar interpretations of the relationship between Islamic and political authority, I rely largely on the oral narratives and commentaries of my Makassar interlocutors and on the detailed chronicles, genealogies, and diaries that Makassar and Bugis royal houses kept in order to document their military, administrative, and diplomatic accomplishments. Like secular historical narratives, the sacred narratives discussed in this book find meaning in particular sequences of historical events, but like myths they interpret them in relation to a universal narrative of transcendental truth. At one level, these narratives purport to be historical accounts of unique individuals who lived at a certain time and place. At another level, they have a universal and unchanging significance for all members of a religious community. They are recited again and again during collective rituals tied to calendrical cycles of varying lengths, and to the human life cycle. In this capacity they have many of the same qualities as myths and periodic rituals that situate particular individuals to an unchanging transcendental order (see Pelras 1979 for a general discussion of Bugis and Makassar narratives; compare the discussion of Filipino narratives in Ileto 1979 and Rafael 1988). In the course of my analysis, I introduce the concept of "symbolic work" to describe the way these narratives try to knit together the many different types of symbolic knowledge and experience that Makassar are exposed to in their everyday lives.

In the world religions, the unique life experiences of the founding prophet do more than provide raw material for scriptural narratives. They also provide a template for calendrical and life-cycle rituals that train individuals to model their lives on that of the prophet. The most complete fusion of the self and the prophet is achieved by those who master ascetic disciplines that enable them to achieve a state of transcendental consciousness similar to the one that was granted to the prophet. Where the meanings embedded in *traditional* rituals work their way up into an individual's consciousness and are given verbal expression in the form of myths, religious rituals work the abstract explicit meanings contained in scriptural narratives down into an individual's unconscious.

The universal forms of individual experience and political authority that are based on prophetic narratives and religious rituals transcend the forms of knowledge and authority that are based on local oral narratives and traditional rituals. Religious definitions of persons as ethical individuals responsible for their actions in the afterlife transcend traditional definitions in terms of specific social roles, and religious definitions of rulers as responsible for defending scripturally sanctioned doctrines and practices transcend traditional definitions of dynastic authority. These eschatological considerations create a rupture between the aspect of each individual that is defined in relation to the local social group and the aspect that is called upon to exercise a degree of ethical autonomy within the framework of the world religion.

In the premodern Islamic world, civilizational unity was maintained by a constant outflow of peripatetic merchants and mystics from the sacred center to the periphery where they settled and married local women; by a constant inflow from the periphery to the sacred center of those in search of higher forms of religious knowledge; and by the cosmopolitan networks of religious scholars and mystics that resulted from these flows. The existence of a fixed set of sacred scriptures and of cosmopolitan flows of religious experts, verbal commentaries, and ritual techniques meant that every locality in the Islamic world had access to political and religious models that originated in remote times and places. If Edmund Leach was right that the Kachin of highland Burma had access to three different models of the political order (Shan Buddhism, *gumsa* hierarchy, and *gumlao* equality), Muslims have access to many more (Leach 1954). This makes nonsense of the notion that all Muslims operate according to a single, coherent "personality system" or that they are associated with a single, coherent "cultural system." The study of the scriptures and the performance of the pilgrimage have always supplied them with a wealth of competing models of self and society from which to choose.

Colonial Knowledge and Bureaucratic Authority

The third symbolic complex I discuss in this book is based on what I call documentary knowledge and bureaucratic authority. The genealogy of this complex goes back to the great "hydraulic empires" of ancient times, which relied on an archive of explicit, quantitatively precise documents containing information for the use of a class of impersonal office holders (Wittfogel 1957). This complex was significantly expanded in the early modern era as new military technologies led to the emergence of "gunpowder empires" in many parts of Eurasia, including the absolutist monarchies of Western Europe and Russia, and the Ottoman and Mughal Empires of the Islamic world (MacNeill 1982). It only became the dominant symbolic complex in most social formations with the introduction of universal schooling in the nineteenth and twentieth centuries.

The power of premodern bureaucratic states rested in part on the superior military training of their soldiers and in part on the patient accumulation of impersonal, objective documentation written by bureaucratic officials who were regularly rotated from one post to another. Documents take the process of uncoupling knowledge from particular individuals, times, and places one step further than do sacred narratives. They are composed by self-conscious economic or political agents whose purpose is to convey as much useful information as possible to other agents in the same impersonal organization. Utility is narrowly defined in terms of maximizing the organization's wealth or power.

Weber argued that modern corporations and state bureaucracies take this type of knowledge and authority a step farther. Under fully developed capitalism, a certain kind of autonomous, bourgeois self is generated in everyday life by the constant necessity to make "rational choices" among competing occupations and commodities. It is generated in an even more explicit form in institutions such as schools, corporations, and government bureaucracies where individuals are under constant pressure to demonstrate their superior abilities to produce and consume impersonal information (Weber 1978: 225).

Bureaucratic knowledge and power was introduced into South Sulawesi by the VOC. The use of bureaucratic methods to achieve capitalist ends in large-scale corporations represented a significant advance in the rationalization of power. As Chomsky has argued, the genealogy of the dominant institution of our time, the transnational corporation, can be traced back not to the free enterprise celebrated by Adam Smith but to the ruthless pursuit of wealth and power by the monopolistic Companies against which he fought (Chomsky 1993). The superior military and organizational capabilities of the VOC enabled it to impose humiliating treaties on kingdoms throughout Island Southeast Asia during the seventeenth century. The traditional authority of local rulers and the charismatic authority of cosmopolitan *shaikhs* were called into question by the political and economic success of the VOC. During the eighteenth century, Indonesian Muslims developed three very different attitudes toward the power of the VOC, which varied according to their local political situation. In kingdoms that remained independent of and hostile to the VOC, local sultans turned to cosmopolitan sources of charismatic knowledge and authority. *Ulama* and *shaikhs* who had studied in the holy cities of Mecca and Medina were invited to take noble wives and to implement the Islamic laws then prevalent in the Ottoman Empire. In areas that fell under the direct control of VOC officials, many people turned to otherworldly mystical practices that prepared them for martyrdom in the battle against their infidel overlords. In kingdoms whose rulers allied themselves with the VOC to bring down their traditional rivals, the VOC was portrayed as a morally neutral force.

During the nineteenth century, the bureaucratic methods used within mercantilist enterprises such as the VOC were increasingly applied to the administration of entire societies. Colonial states sought to create the conditions for capitalist economic development by replacing feudal forms of land tenure and service with markets in land and labor. They sought to undermine the traditional authority of hereditary rulers by replacing them with salaried civil servants. They also sought to impose strict limits on the charismatic authority of local religious experts by relegating religion to a private sphere.

These attempts by metropolitan governments to transform the religious, political, and economic systems of their colonies required the accumulation of an unprecedented amount of documentary information on local societies by ever-growing civil and military bureaucracies. This transformation of social life was accomplished through the modern disciplinary techniques outlined by Foucault. These techniques organized bureaucratic space and time according to an abstract plan that allowed standardized units to be subdivided, combined, substituted, and manipulated in an endless variety of ways. Disciplinary enclosures such as asylums, hospitals, barracks, schools, and prisons enabled bureaucratic subjects to gather systematic knowledge about objectified individuals that could be used to mold them into standardized parts of the organization. These disciplinary techniques were the bureaucratic analogues of the traditional and charismatic rituals discussed earlier. It is no accident that many of them were first developed within the enclosed disciplinary space of the medieval monastery.

As time went on, the personnel required to staff these bureaucracies were increasingly recruited from members of the local population who had been trained in modern schools and military training camps. The colonized people who received this training quickly appropriated the ideas of a secular state composed of free citizens with equal rights and began to demand national autonomy. Indeed, despite the pretensions of the great prophetic religions to universality, nationalism was the first truly global ideology (Anderson 1991).

For much of the twentieth century, political discourse was dominated around the globe by competing versions of nationalism. These included the statist capitalism inspired by Napoleonic France and the free-market capitalism inspired by Victorian Britain; the bureaucratic socialism inspired by Soviet Russia and the peasant-based socialism inspired by Maoist China; and the Islamic modernism inspired by Egypt and the Wahhabi conservatism inspired by Saudi Arabia. All of these competing ideal models were based on the same documentary conception of knowledge as explicit, impersonal, and rational as opposed to the implicit, personal, and intuitive knowledge acquired under the guidance of a Sufi *shaikh*.

Despite the predictions of classical "secularization" theory, the rise of democratic nationalism no more leads to the demise of prophetic religion than the rise of prophetic religion lead to the demise of traditional ritual and myth (Berger 1967). What tends to happen instead when a new symbolic complex is introduced into a social formation is that older complexes undergo a process of specialization and differentiation. Religious systems that once provided both ethical templates for how to live and cognitive models for interpreting the meaning of the cosmos lose much of their cognitive function once the impersonal types of knowledge demanded by bureaucratic institutions become dominant. Religious and bureaucratic selves continue to be generated by distinct symbolic complexes that produce different psychic needs and proclivities.

During times of transition, there will always be those who resist the process by which older symbolic complexes find their scope restricted as newer complexes are introduced. Contemporary examples of this kind of resistance include radical Islamists who would eliminate both nationalism and democracy in the name of a revived caliphate;

Christian advocates of "creation science"; and Hindu advocates of "Vedic science" (Qutb 1981; Toumey 1994; Alter 2004). But as the aging Makassar militants who fought to establish *shariah* law as the sole foundation of the Indonesian nation-state in the 1950s had discovered by the 1980s, it is impossible to reduce all the forms of symbolic knowledge within a complex social formation to a single coherent model. Local social hierarchy, democratic nationalism, and religious cosmopolitanism have come to terms with one another in Indonesia, as they have in most other social formations around the globe (for Indonesia, see Hefner 2000; for Christian parallels, see Casanova 1994).

Outline of the Chapters

I bring this theoretical discussion to a close with a brief overview of the seven ideal models discussed in this book. It is my contention that all these models remain accessible to political actors today. Each provides a potential contrast to all the others, and so helps to define them. Southeast Asian Muslims are thus never at a loss for alternative models against which they can measure the existing political order. The continued legitimacy, or hegemony, of that order has always been open to question. This is not because actors contrast the dominant order with what they know to be their true self-interest, but because the multiple symbolic complexes in which they are embedded make available to them many different kinds of selves and kinds of interest among which to choose. Every ideal model represents a temporary synthesis of all the contradictory experiences of self and society that are being generated by the symbolic complexes present at a particular moment in time and space.

In chapter 2, I examine an ideal model under which traditional, charismatic, and bureaucratic forms of authority are fused in the hands of a single ruler. I begin with the social and political conditions in the ancient Middle East that led to the appearance of ethical prophets and to the codification of their message in the form of sacred scriptures. I discuss how the world empires that later based their claims to universal authority on these prophets eventually evolved into politically decentralized religious commonwealths. I then turn to the history of the relationship between the supreme political and religious authorities in Islam from the time of the Prophet in the seventh century to the rise of the gunpowder empires of the sixteenth century. I show that when the kings of South Sulawesi converted to Islam at the beginning of the seventeenth century, they did so by adopting a doctrine developed during the sixteenth century in Iran and northern India that an Islamic ruler should serve as the mystical link between God and his subjects. This doctrine enabled the early Makassar and Bugis sultans to preserve the exalted status they derived from traditional myths and rituals, which portrayed them as the descendents of local divinities, alongside the new charismatic authority they derived from their mastery of Islamic knowledge.

In chapter 3, I examine an ideal model in which the traditional authority of a local ruler is viewed as complementary to the charismatic authority of cosmopolitan religious experts. This model derived from the politically decentralized lands surrounding the Arabian Sea where a synthesis between the study of the *hadith* and the practice of mysticism developed during the sixteenth and seventeenth centuries. Local rulers in this area typically deferred to the cosmopolitan *ulama* and *shaikhs* who traveled from one city-state to another throughout the region. Pilgrims from Southeast Asia were exposed to this cosmopolitan model of Islam at the Indian and Arabian ports they visited on their way to Mecca. One of the most famous examples of such a charismatic *shaikh* was Yusuf al-Maqasari (1626–1699). Yusuf left Gowa in 1644 and traveled to Mecca by way of Java, Sumatra, India, and Yemen. He returned to Indonesia in about 1670 and became the chief religious adviser in Banten, where he implemented the strict form of Islamic law that was then current in the Ottoman Empire. One of his students, Tuan Rappang, implemented similar policies in Gowa. Seeing his cosmopolitan charisma as a threat, the VOC captured Yusuf in 1683 and exiled him to Sri Lanka. He was then moved to South Africa in 1693, where he died in 1699. Yusuf's body was returned posthumously to Gowa in 1705, where it was buried in state near the graveyard of the kings. According to a traditional account of his life, Yusuf returned from death to impregnate the sultan's daughter and to establish a line of charismatic *shaikhs* and a local branch of the Khalwati Sufi order with whom the royal house of Gowa established a relationship that endured into the twentieth century.

In chapter 4, I examine an ideal model in which the traditional authority of local rulers, the charismatic authority of mystical adepts, and the bureaucratic authority of the VOC are seen as absolutely incompatible. This model developed in the territories that became vassals of the VOC in 1667. Many Makassar subjects of the VOC found it impossible to reconcile themselves to being ruled by foreign infidels who made most of their personal and official profits out of the slave trade. Their reaction to Dutch rule and to the corrupt local lords who collaborated with them is preserved in the *Epic of Datu Museng*, an oral epic about a minor noble from Sumbawa who died in single combat with VOC troops in 1767. This epic continued to evolve throughout the nineteenth and twentieth centuries, acquiring new meanings as the political context changed. By 1989, it had come to serve as an allegory for the human life cycle, from birth through marriage to death; for the history of the Indonesian people from the formation of local kingdoms, through their conversion to Islam, to their subjugation by the VOC; and for the mystical path followed by all creation from the beginning to the end of time.

In chapter 5, I examine an ideal model in which the relationships between the traditional authority of local rulers, the charismatic authority of cosmopolitan *shaikhs*, and the bureaucratic authority of the colonial state are regarded as complementary. This model developed during the nineteenth century when the remaining traditional authority of the royal houses of Bone and Gowa had been consolidated through several generations of intermarriage; charismatic authority had shifted from the *ulama* and *shaikhs* patronized by the royal courts to popular Sufi orders such as the Sammaniyya; and the local population had come to accept the bureaucratic authority of the colonial state over secular matters. This model is reflected in *Epic of the Three Boats*, which recounts the fall of Gowa in 1667 to the alliance between the VOC and Arung Palakka, a Bugis noble from the kingdom Bone. In this epic, the VOC is portrayed as a morally neutral instrument of divine wrath against an impious king. Arung Palakka is portrayed as the son of the king of Gowa who is forced by his father to flee the kingdom. His exile and triumphant return are modeled on the life of the Prophet Muhammad as told in the *Maulid al-Nabi* by Jaffar al-Barzanji (d. 1767). Like the spread of the Sammaniyya, the popularity of al-Barzanji's *Maulid* throughout the Islamic world in the nineteenth century was indicative of the new sense of mystical devotion to the spirit of the Prophet rather than to Sufi *shaikhs*.

In chapter 6, I examine an ideal model based on the complementarity that developed between colonial officials and the growing number of Muslim villagers who were able to perform the *hajj* between 1850 and 1950 due to the introduction of steamships. For much of this period, these two groups regarded the traditional power and authority of the old noble families as an obstacle to economic and spiritual progress. I explore these relationships through a discussion of oral histories I collected in the village of Ara about Panre Abeng, a commoner who moved to Ara in 1890, and about four generations of his descendents. Panre Abeng's son, Haji Gama served as village head of Ara from 1915 to 1949, and is remembered for rigorously enforcing *shariah* law while loyally serving the colonial state. Since the colonial administration enforced a strict separation between political and religious affairs, reformist impulses were channeled into a critique of the un-Islamic nature of local ritual practices rather than of the un-Islamic nature of the state. Haji Gama's descendents transferred their allegiance to the secular state of the Republic of Indonesia, and many went on to achieve high office in the provincial government. Many used their political and economic success to marry into traditional noble families, and they eventually made their peace with the traditional ritual system through which noble rank was reproduced.

In chapter 7, I examine an ideal model in which the distinction between charismatic religious knowledge and documentary secular knowledge is collapsed, while all forms of traditional symbolic knowledge are rejected as un-Islamic. The roots of this model lay in the schools set up by colonial states in India, Egypt, and Indonesia to produce a cadre of low-level bureaucrats. Widespread literacy had the unintended effect of producing a market for printed literature that included both nationalist critiques of the colonial state and modernist critiques of traditional Islamic practices. In 1952, the supporters of the Darul Islam movement began a guerilla war against the secular government. They suppressed all manifestations of the traditional cults of the royal ancestor spirits and of the charismatic cults of the village *shaikhs*. Government authority was finally restored by 1965, but the Darul Islam militants continued to actively campaign against the cults of the royal ancestors and the *shaikhs*.

In chapter 8, I show that President Suharto deliberately manipulated many of the ideal models of Islamic piety and political authority discussed in earlier chapters. Suharto claimed traditional authority by stressing his ties to the royal courts of central Java. He claimed charismatic authority by portraying himself as a lowly orphan who had acquired a noble wife and political power through his military prowess, in a manner that recalled the *Epic of Datu Museng*. He also claimed to be a master of esoteric knowledge like Shaikh Yusuf, an orthodox *hajji* like Haji Gama, and an Islamic nationalist like the leader of the Darul Islam movement. Above all, however, he emulated the centralized bureaucratic authority of the Dutch colonial governor generals. In 2000, it was still far too early to tell how much of this version of his life would be accepted by future generations of Makassar. A solid majority of South Sulawesi voted for Suharto's party, Golkar, in the 1999 elections, only to abandon it in 2004.

In the conclusion, I return to the question of the relationship between the explicit, verbal forms of symbolic knowledge discussed in this book, and the implicit, sensual forms of knowledge, such as the life-cycle rituals and ascetic disciplines that I will discuss in a future book. I do so to remind the reader that the symbolic models I analyze here are linked to the altered states of consciousness engendered by ritual performances and mystical disciplines. All these models are still very much alive and continue to provide Makassar individuals with a range of alternative ways of conceiving both the external world of society, religion, and politics and the internal world of the relational person, the ethical soul, and the self-interested actor.

Chapter 2

The Ruler as Perfect Man in Southeast Asia, 1500–1667

Introduction

In this chapter, I discuss the reasons for the conversion of almost all the rulers of South Sulawesi to Islam in the remarkably short period between 1605 and 1611. I begin by going back to the very origins of ethical prophecy, charismatic authority, and alphabetic scriptures among the urban artisans and merchants of the ancient Middle East. I argue that their exposure to the competing symbolic traditions of the powerful agrarian empires of Egypt, Mesopotamia, Iran, Greece, and Rome tended to undermine their confidence in the validity of their own symbolic traditions and made them receptive to religious visions with a more cosmopolitan relevance. These cosmopolitan symbolic systems also proved attractive to political rulers who were trying to govern multiethnic world empires. I show that similar forces were at work in Island Southeast Asia between 1300 and 1600, and this is what explains the conversion of the entire region during this period.

A number of different models developed within the early Islamic commonwealth concerning the proper relationship between the supreme political authority and the religious authorities. These ranged from the early caliphs who claimed supreme political and religious authority over the entire Islamic world, to the military rulers of small city-states who were happy to acknowledge the charismatic religious authority of cosmopolitan *ulama* and *shaikhs* who traveled from one court to another in search of masters and disciples. In the sixteenth century, there was a brief revival by Shah Ismail of Iran, Emperor Akbar of India, and Sultan Agung of Java of the old caliphal model in which the ruler claimed supreme charismatic authority for himself as the exemplar of Ibn al-Arabi's concept of "the Perfect Man." It was this revival that paved the way for the conversion of the kings of South Sulawesi at the beginning of the seventeenth century.

Prophecy and Charisma

According to Weber, ethical prophecy emerged in the ancient Middle East among the less developed peoples of the region who "tended to see in their own continuous peril from the pitiless bellicosity of terrible nations the anger and grace of a heavenly king" (Weber 1963: 59). They developed a peculiarly explicit consciousness of their place in world history because they existed on the margins of the "relatively contiguous great centers of rigid organization" constituted by the agrarian empires of Egypt, Mesopotamia, Iran, Greece, and Rome. The coexistence of profoundly different symbolic systems in the ancient Middle East led to a generalized decay in the taken-for-granted validity of traditional religious and ritual practices, especially in the urban centers. As they grew disenchanted with their particularistic tribal gods and myths, the geographically and socially marginal people of the region became receptive to the increasingly universal claims made by charismatic prophets.

The universal norms articulated by ethical prophets appealed primarily to artisans and merchants whose way of life was responsive to rational calculation and self-discipline. By contrast, rural peasants who were dependent on uncontrollable natural forces remained fundamentally magical in their approach to religion (Weber 1963: 97–98). Jesus and Muhammad were close to the ideal type of ethical prophet for Weber. Both belonged to peoples threatened by the "pitiless bellicosity of terrible nations." Both belonged to the middle classes: Jesus was a carpenter and Muhammad was a merchant. Finally, both stood "at the lower end of or outside the social hierarchy." Jesus had no socially recognized biological father and Muhammad was an orphan. Their power was not based on traditional rules of patriarchal kinship but on their charismatic connection to a completely different divine order. But neither were they completely outside the traditional social hierarchy. Joseph, the legal father of Jesus, was descended from King David (Luke 1: 26–33). Muhammad was a member of the powerful Quraysh tribe. They thus stood at the "point of Archimedes" in their respective societies.

Weber sometimes used the term "charisma" to describe the particular kind of knowledge and authority wielded by prophets as opposed to priests or bureaucrats.

We shall understand "prophet" to mean a purely individual bearer of charisma, who by virtue of his mission proclaims a religious doctrine or divine commandment . . . [The] personal call is the decisive element distinguishing the prophet from the priest. The latter lays claim to authority by virtue of his service in a sacred tradition, while the prophet's claim is based on personal revelation and charisma. (Weber 1963: 46)

Benedict Anderson has suggested that Weber's concept of "charisma" only makes sense when it is linked this way to prophetic religions.

These religions are, or were, "first class," World religions, because, so it seemed to him, they were conceived in terms which, *in principle*, made them open to all human beings. The advent of these religions—Buddhism, Hinduism, Judaism, Christianity, Islam, and Confucianism—struck him as absolutely astonishing and revolutionary, and as (quite often) the product of "extraordinary" individuals—Gautama, Jesus Christ, Muhammad—who by their visions brought about radical breaks in the civilizations into which they had been born. (Anderson 1985: 80)

Anderson goes on to note, however, that Weber ended up conflating the charismatic innovations of the prophets with the quite different idea that hereditary war chiefs and kings possessed extraordinary magical powers whose properties were strictly defined by tradition (Weber 1963: 2).

While I agree with Anderson's view that Weber's use of the term "charisma" is confused, I think there are many advantages to retaining the concept if its meaning is restricted to the universal knowledge that derives from prophetic revelation. It may then be contrasted both with the traditional knowledge embedded in anonymous myths and rituals as well as with the "rational" knowledge and power contained in impersonal bureaucratic documents. Charismatic knowledge and power derive from a divine realm governed by rules that transcend any particular social or political order. Charismatic knowledge often confers mystical powers that also transcend ordinary laws of time, space, and causality.

In the premodern world, this kind of charismatic power could represent a threat to traditional and to bureaucratic political authority. The universal laws enunciated by the prophets could be interpreted in such a radically populist way that all political authority was delegitimated. They could also be interpreted in such an elitist way that only a few "religious virtuosi" could truly observe them. On the basis of their superior transcendental knowledge, religious elites could claim to outrank not only ordinary people but traditional social and political elites as well. Traditional rulers could try to appropriate charismatic authority for themselves by making religious elites dependent on them for patronage. But in a world of many political centers, religious elites could always seek alternative patrons. In the early modern world, many absolutist states tried to assert "caesaro-papist" control of the religious hierarchies in their lands, but such attempts lasted little more than a century. More recently, many postcolonial states have attempted to harness charismatic power by imposing official interpretations on religion in the schools and mass media. But popular interpretations of such official teachings have usually proved impossible to control.

Scriptural Populism and Philosophical Elitism

Sacred scriptures are prophetic narratives that have been written down in a form that allows them to maintain their integrity over long periods of time. Prophetic narratives are fixed at the moment they are recorded, but the messages they convey are universal enough to remain intelligible as they travel through time and space. For this to be possible, these messages must be far more abstract and explicit than those conveyed by rituals and myths that can be continually modified to fit the current situation. In this they are similar to philosophical texts that also try to transcend the limitations imposed by local cultural tradition and to achieve a level of universal truth. One of the differences between prophecy and philosophy lies in the popular appeal of prophecy, which addresses itself to questions that vex ordinary people in their everyday lives. For prophetic teachings to become accessible to a large popular audience, however, they had to be recorded in a medium that was relatively easy to decode. This was provided by the development of alphabetic scripts "in an area situated between the early written civilizations of Egypt and Mesopotamia among a people known as the Canaanites, the Semitic-speaking inhabitants of Syria and Palestine before the coming of the Israelites from whom they are difficult to distinguish . . . [This was a] region of small kingdoms and rich merchant princes" and "the meeting place of invaders and cultural influences not only from Egypt and Mesopotamia, but also from the north where the Hurrians and their Mitanni rulers, probably originating in central Asia, spoke an Indo-European language" (Goody 1987: 43). The alphabet was thus the product of a cosmopolitan, mercantile, polyglot, and urbanized world. The spread of alphabetic writing went hand in hand with the spread of a common language for the conduct of trade and diplomacy.

It is no coincidence that religions based on the popular visions of prophets arose at precisely the same time and place as alphabetic writing. The alphabet, like prophetic knowledge, presents a challenge to the monopoly of priestly elites over sacred knowledge. Their combination in religious scriptures that are accessible to the urban middle classes had revolutionary implications for the distribution of symbolic knowledge in the ancient world.

Marshall Hodgson argued that the prophetic religions of the ancient Middle East were intrinsically populist, in line with their origin among the literate middle classes (Hodgson 1974 I: 130). But however populistic a world religion might be, when it became the shared symbolic system of an entire social formation it also had to develop an elite version that could appeal to the upper reaches of the social hierarchy. In the ancient Middle East, elite forms of religion tended to draw on Greek philosophy in a way that complicated the straightforward message of the ethical prophets. Like ethical prophecy, the beginnings of Greek philosophy may be traced to a growing recognition that the literal truth of traditional mythological systems was no longer tenable in a cosmopolitan environment. But unlike the Semites of the eastern Mediterranean, the Greeks experienced Middle Eastern history as a conquering elite. Their philosophy was not meant to explain the victimization of marginal peoples by vastly more powerful forces, but why their institutions were superior to those of the barbarians (Hodgson 1974 I: 411). It was only after the first generations of Hellenized converts to Christianity and Islam succeeded in combining Greek philosophy with Semitic prophecy that they could become truly universal in their appeal to both ruling elites and popular classes. The juxtaposition of philosophical reason and prophetic revelation proved to be a source of endless creativity not only for these religions, but for the Judaic tradition as well.

World Religion and World Empire

Scriptural religions based on prophetic revelation transcended the particularities of time, place, and cultural tradition. This made them better suited to serve as unifying ideologies for the empires of the ancient world than did the particular symbolic traditions of whichever society first established the empire in question. Perhaps the earliest example of a ruler who adopted a prophetic religion to unify his empire was Ashoka (ca. 272–231 BCE), who converted to

Buddhism two centuries or so after the death of its founder, Gautama (Tambiah 1976).

Christianity and Islam both developed as a result of a long-term dialectic between tendencies toward political and cultural unification and fragmentation in an area that stretched from western Asia through the Mediterranean basin. World empires created the conditions for the spread of world religions, while world religions both legitimated and outlived world empires.

> The Islamic Empire was actually and aggressively universal, Islam only potentially—and the empire consented to tolerate a degree of cultural-religious pluralism. Had the Islamic Empire been prepared to tolerate only Islam, it would have had to impose the inhuman uniformity for whose sake Constantinople had vainly struggled through more than three centuries of Christological debate. It would have dissipated its energies in internal strife, and "might well have shrunk back to the wastes of Arabia from which it had sprung." Alternatively, Islam would have become the very diverse religion it eventually became anyway, but without the memory of the Golden Age of the Abassid Baghdad—one god, one empire, one emperor—to sustain it. . . . The resulting relation-ship between world empires, world religions and cultural commonwealths was close but indirect. (Fowden 1993: 160, 169)

Caliphal Elitism, 632–848

During the period of transition from a local Arab religion to a world religion encompassing many ethno-linguistic groups, techno-environmental adaptations, and socioeconomic strata, Islam developed many disparate tendencies. One of these, the assignment of a central religious function to the ruler, was crucial for the spread of Islam into the centralized, hierarchical societies of monsoon Asia. From the very beginning of Islam, a part of the community assigned Muhammad's successors, the caliphs, a central place in the regulation of both reli-gion and the state. Crone and Hinds have argued that for the first two centuries after the death of Muhammad, the caliphs saw themselves as literally the Deputies of God on Earth with a status almost equal to that of the prophets. In this respect, they argue that the Shia tradition of the imams as the genealogical and religious as well as the political successors of Muhammad may well be closer to the original form of the caliphate than the later Sunni view of the caliph as merely Muhammad's political heir (Crone and Hinds 1986).

The status of the caliph was further exalted when the Abbasids moved the capital of the Islamic Empire from Damascus to Baghdad in 762 CE. Baghdad was located at the heart of Mesopotamia, which had long served as the granary of the Sassanian Empire. The Abbasid caliphs began to model themselves on the Sassanian kings, and added the Persian title of "The Shadow of God on Earth" to that of "Deputy of God." State revenues came to depend on the taxation of peasants working irrigated land, and the caliphs became much like the absolute rulers of previous agrarian empires in the Fertile Crescent. Between 813 and 848, the Abbasid caliphs patronized the Mu'tazila, an elitist philosophical school that tried to reconcile Muhammad's revelation with Greek rationalism. They held that the truths of Islam were accessible through the exercise of individual reason, and downplayed the necessity of prophetic revelation.

Scriptural Populism, 848–1171

The Mu'tazila philosophers were opposed by a populist movement known as the *Ahl al-Hadith*, the *Hadith* Folk. The *hadith* were collections of texts reporting the words and deeds of the Prophet Muhammad when he was not transmitting divine revelations but which were nevertheless considered to be a secondary source of guidance in the conduct of human affairs. The *Ahl al-Hadith* held that individual reason could never achieve the divine knowledge given to the prophets. Only the Koran and *hadith* could be used as a guide to correct belief and action. The Koran became the focus of such veneration that it was held to be an "uncreated" aspect of God. Religious primacy thus belonged solely to the *ulama*, "The Learned Ones," those schooled in the interpretation of these fixed texts. Human reason must always be the servant of the texts. This movement was associated with the development of the classical Sunni schools of law, all of which formed during the Abbasid period.

These views left little or no religious role for the Sunni caliph, the Shia imam, or indeed any intellectual elites aside from the *ulama* to play. The claims of the *ulama* thus posed a challenge to the religious pretensions of the caliphs. The *Ahl al-Hadith* were subjected to severe repression under the caliphs al-Ma'mun (sole ruler 813–833), al-Mutasim (833–842), and al-Wathiq (842–847). The last and most militant of the schools of law based on the Koran and *hadith* was that founded by Ibn Hanbal (780–855) at the end of this period of caliphal repression. He was condemned to death by al-Ma'mun and was saved only when the latter died before the sentence could be carried out.

Ibn Hanbal lived to see a complete reversal of fortunes of the *Ahl al-Hadith*. In 848, the Caliph al-Mutawakkil (847–861) ended their persecution, and they soon came to define Islamic orthodoxy. The caliphate itself went into accelerating decline after this. Caliphal extravagance, a decline in agricultural revenues, and a growing reliance on Turkish military slaves combined to weaken absolutist claims to power, while a boom in trade with both Tang China and Europe increased the wealth of the merchant class. After 850, real power lay in the hands of Turkish pastoralists and real wealth in the hands of urban merchants. This new arrangement was finally rationalized in the twelfth century by *ulama* such as Abu Hamid al-Ghazali (d. 1111). Al-Ghazali's understanding of the relation between caliph, sultan, and *ulama* became general in the Middle East when Fatimid rule in Egypt was brought to an end by a Kurdish Sunni, Salah al-Din, in 1171.

Mystical Elitism, 850–1250

Philosophically inspired interpretations of Islam may have been marginalized by the *Ahl al-Hadith*, but they did not disappear. Beginning with al-Kindi (d. ca. 870), certain intellectuals began to try to reconcile Muhammad's divine revelation with rationalist Greek cosmology as found in the works of the Neoplatonists (Arberry 1957). In the tenth century Neoplatonism became popular in Ismaili Shia circles, first in Persia and then under the Fatimids in Egypt. In the eleventh and twelfth centuries, this synthesis reached its apogee with the Persian Ibn Sina (Avicenna, d. 1037) and the Andalusian Ibn Rushd (Averroes, d. 1198). In highly abstract language, these Islamic philosophers described the phenomenal world as the product of successive emanations from God's Unitary Being. Revealed religion was considered to be but a "philosophy for the masses," expressing the truths of philosophy in imaginative symbols that served their "moral edification and purification" (Rahman 1979: 118).

The rationalist and elitist approach of these philosophers was soon rejected by the Islamic mainstream. Many of their insights were rescued, however, by Ibn al-Arabi (1165–1240) who translated their basic concepts from the plane of abstract rational argumentation to that of immediate mystical experience. Where earlier mystics had resorted primarily to poetic imagery to express their experiences of unification with God, Ibn al-Arabi adapted the complex conceptual system of Neoplatonic philosophy to express his immediate mystical experiences. The truth of his system was grounded not in individual reason alone but also in the intuitive knowledge bestowed on the devout mystic by God. The speculative metaphysics of late antiquity acquired in this way an existential and ontological reality in Islamic civilization.

Neoplatonism proved more acceptable to the Islamic mainstream when it was seen as the product of divinely guided intuition than as the product of human reason. According to this view, God's creation of the world was not an arbitrary act. It was motivated by His desire to be both subject and object of knowledge. The telos of Man is to reascend the levels of creation to realize the essential unity of all that exists, to truly know God in the way He intended to be known. He who does so is a Perfect Man. The universe was thus created so that there might be a Perfect Man, and the Perfect Man was created so that God might know Himself. In the Age of the Prophets, from Adam to Muhammad, God granted knowledge of Himself in each generation to a particular man. These men were simultaneously *shaikhs* who were given insight into the inner nature of reality, and prophets who were given a set of laws governing the external nature of reality. Muhammad was the most perfect of the prophets and of the *shaikhs*.

The enormous corpus of Ibn al-Arabi's writings went on to inspire generations of esoteric mystics. The corpus was so voluminous and difficult that its popular impact mostly occurred by way of later interpreters. One of the most important of these was Abd al-Karim al-Jili (d. 1428). He claimed descent from Abd al-Qadir Jilani and his thought made a particular impact on members of the Sufi order that followed his teachings, the Qadiriyya. Al-Jili clarified Ibn al-Arabi's basic insight regarding the ultimate unity of being (*wahdat al-wujud*) in terms of a series of emanations or Grades of Being. The main problem was to reconcile God's ultimate unity as experienced by the mystic with the phenomenal multiplicity of the world as experienced in everyday life.

One might say that among Ibn al-Arabi and his followers, mysticism achieved an ultimate form of elitism. According to the doctrines of the *Qutb* and the *Insan al-Kamil*, a single human being serves as the linchpin for the whole of creation. This elitism is universalistic and potentially in tension with all local social and political hierarchies, although it is at least potentially compatible with them. As we will see, a temporary fusion of the mystical and political hierarchies was to take place not in the Arab heartland of Islam, but in Iran and northern India.

Mystical Populism and Techniques of the Body 1100–1328

For several centuries, the pursuit of mystical experience remained confined to a dedicated elite. Popular methods for achieving immanent experience of the transcendental truths of Islam were only gradually perfected. Muslim mystics first borrowed certain disciplines from the Christian monks of the Middle East, such as prolonged withdrawal from the world, fasting, meditation, and perhaps even the rhythmic repetition of formulae designed to induce the "remembrance" of God, or *dhikr*. Scriptural knowledge came to be accompanied by a whole set of ritual techniques and disciplines that embedded this knowledge in people's bodies (Trimingham 1971: 198).

> The Sufis honoured the Qur'an as enshrining God's message to Muhammad, but rather than devote themselves to the letter of its words, they hoped in some measure to repeat in their own lives something of the experiences which presumably Muhammad must have gone through in receiving various portions of the words of God. (Hodgson 1974 I: 394)

Training in the mystical sciences was not unlike apprenticeship in a craft or training in the martial arts. Craft guilds developed in the Islamic societies of the Middle Ages. Ancestry predisposed individuals to the acquisition of certain trades, but did not rigidly determine them. Significantly, these guilds were also the earliest mystical brotherhoods (Hodgson 1974 II: 221).

Ironically, Sufism proved especially popular among the spiritual descendents of a strict Hanbali ascetic, Abd al-Qadir Jilani (1077–1166). Abd al-Qadir began his career as a rigid Hanbali jurist, the party that had done so much in the ninth century to cut the pretensions of the caliphs and the Mu'tazila philosophers down to size, and ended it as a popular Sufi *shaikh*. There proved to be a fit between a form of populist mysticism and a form of populist scripturalism that made a kind of prophetic experience available to ordinary worshippers. Both denied the claims of the elite to approach more closely to God through the exercise of their superior intellect alone.

His own preaching had been a rather modest call to piety, but after Abd al-Qadir's death in 1166 a cult devoted to his person developed among the masses whose approach to religion was rather more practical than that of religious virtuosi. His followers were among the first to establish a regular *tariqa'*, path to enlightenment. The elite Neoplatonic theosophy of Ibn al-Arabi entered the Islamic mainstream through this *tariqa'* and others like it. Abd al-Qadir's charismatic power was institutionalized by his descendents, who cared for his tomb and for the endowments accumulated in his name. Miraculous tales concerning his abilities in life and his powers as an intercessor in death began to circulate, and he became the most universally popular *shaikh* in Islam (Trimingham 1971: 43).

Charismatic Shaikhs and Military Sultans in India, 1221–1325

The equilibrium between Turkic war craft, Iranian statecraft, and Arabic religious knowledge that had been achieved by the twelfth century was abruptly broken by the Mongol invasions of the thirteenth century. Sufi *tariqa'* played a crucial role in the survival of Islam during the Mongol conquest of the Islamic heartlands of Iraq and Iran. Some authors saw this invasion as threatening a reversion to *jahiliyya*, the state of ignorance in which the Arabs had existed before the coming of the Prophet Muhammad. The most famous of these writers was Ibn Taimiya (1263–1328), who was born just five years after the fall of Baghdad (Makdisi 1974: 129).

[After] 1258/656, most Sunni-Jama'is lived under military dictators whose authority could not be recognized by a pro forma caliphal conferral of legitimacy and whose support of the Sacred Law could not be confirmed by a pledge of allegiance to the caliph, the personification of the preeminence of that Law. Consequently, various forms of sacral sovereignty and legitimacy were the only means available to post-Mongol Muslim warlords to endow their regimes with the trappings of legality. (Woods 1999: 4)

In much of the Islamic world it was the *shaikhs* of the great Sufi orders who became the most important sources of charismatic authority. Many such *shaikhs* sought refuge from the Mongols in the Islamic sultanates of India. In India, these *shaikhs* came to exercise spiritual power over a *wilayat*, or territorial domain. A local sultan whose jurisdiction overlapped with this *wilayat* had to develop a relationship with the local *shaikh* (Digby 1986; Eaton 1978, 1993; Ernst 1992).

The Perfect Man as Ruler in Iran and India, 1300–1602

The Safaviyya was founded in Azerbaijan as a militant Sufi order by Safiya al-Din (1249–1334). One of his descendents, Ismail (1484–1524), took control of the order when he turned sixteen in 1500 and set about conquering much of Azerbaijan, Iraq, and Iran.

He then imposed a mystical version of Shiism on the entire population in which he played a central role (Woods 1999).

> In the first decade or so of Ismail's reign, the state was founded firmly upon the position of Ismail as *murshid-e kabil*, the "perfect [Sufi] master," assisted by his subordinate pirs and khalifahs, to whom were devoted the loyal Turkic tribes of murids, the Kizilbash, who were fulfilling their spiritual discipline in following his military commands. . . . His own verses proclaim him a locus of Divinity for his times, as a descendent of the Twelver imams; and honor the imams themselves in ways that more cautious Shiis could find exaggerated. (Hodgson 1974 II: 31)

Some of these ideas were communicated to northern India by way of the early Mughal rulers. During the second half of the sixteenth century Akbar centralized Mughal power over the agrarian heartland of the Punjab and the Ganges plain using the latest gunpowder technology. He conquered Gujarat in 1573, Bihar and Bengal in 1576, and Ahmadnagar in 1600 (Hodgson 1974 III: 63). As his power grew, Akbar felt he no longer needed the Sufi *shaikhs* to lend him their authority.

> Finally, in 1579, Akbar assumed sweeping powers in matters of Islamic doctrine. An imperial edict publicly stated the Mughal emperor's prerogative to be the supreme arbiter of religious affairs within his realm— above the body of Muslim religious scholars and jurists. . . . The edict also sought to claim for Akbar the authority as *Khalifa* in preference to the Ottoman Sultan who had claimed that title since seizing control of the Holy Cities in 1517. (Richards 1993: 39–40)

Akbar's greatest promoter was Abu al-Fazl, who composed the *Akbar Nama* between 1596 and 1602, a chronicle of the Emperor's reign. Abu al-Fazl explicitly argued that Akbar was the equal of the greatest prophets and *shaikhs* of Islam (Rizvi 1975: 358). With Akbar, the Islamic theory of the state seemed to have come full circle, with the same individual enjoying supreme religious and political power, and it was this model that finally persuaded the rulers of South Sulawesi to join the Islamic commonwealth.

The Ruler as Perfect Man
in Southeast Asia 1500–1667

Until their conversion to Islam, the rulers of kingdoms in the lands surrounding the Java Sea legitimated their authority in terms of a symbolic language derived from a blend of ancient Austronesian and Indic myths and rituals (Gibson 2005). Hindus from the Chola kingdom of South India played a prominent role in maritime trade with Southeast Asia until it collapsed in the thirteenth century. They were replaced by Muslims from north India, who taught a form of Islam in which the traditional authority of local rulers was combined with the charismatic authority of cosmopolitan Sufi *shaikhs*. The combination of Islamic hegemony over the sea-lanes and a form of Islam that reinforced the authority of existing royal houses proved irresistible to local rulers, and by 1600 almost all the rulers around the Java Sea had converted to Islam.

The kings of South Sulawesi were among the last to convert. They were only persuaded to do so when a version of Islam became available that placed the ruler himself at the apex of the religious hierarchy. Between 1605 and 1611, they adopted the model of Islamic kingship that was then fashionable in the great "gunpowder empires" of Iran, India, Aceh, and central Java. It was relatively easy for them to transform the existing Indo-Austronesian model of the king as descendent of the divine ancestors into the Islamic model of the king as the Perfect Man. The cosmopolitan narrative of Islam was in this way appropriated by the rulers of South Sulawesi and transformed so that it could be integrated into the existing regional narrative that legitimated their political authority. Following the same logic, the cosmopolitan Islamic idea of the *wali Allah*, or friend of God, was appropriated and used to transform regional shrines dedicated to the royal ancestors into the shrines of Islamic *shaikhs*.

The Formation of an Islamic Commonwealth in the Indian Ocean

Among the most active Muslim merchants in Southeast Asia during the fourteenth century were Gujaratis from western India. Gujarat was under the control of the Delhi Sultanate between 1303 and 1407. During this period, it "became the chief importer of the luxury goods demanded by the conspicuously consuming Delhi elite," and Gujarati merchants proliferated in the ports of Southeast Asia (Abu Lughod 1989: 272). One of their first stops was the ancient kingdom of Barus on the western coast of Sumatra. During the fourteenth century, "Barus was not only an important cosmopolitan commercial center, as was already the case previously, but it had become one of the Islamic centers of Indonesia, thanks to the masters of religion (*shaikh*), come without doubt from abroad, who lived there and taught there and of whom certain tombs still bear witness" (Guillot and Kalus 2000: 22). These masters brought with them the doctrine that the charismatic religious authority of cosmopolitan *shaikhs* was complementary with the traditional political authority of local sultans. A local king need only replace his court Brahmin with a court *shaikh* to achieve legitimacy in the eyes of his fellow Muslim rulers.

The formation of an Islamic commonwealth in Island Southeast Asia received another boost from the expansion of Chinese maritime trade during the first half of the fifteenth century. The third Ming ruler, Yongle, thus commissioned a series of six naval expeditions between 1405 and 1435 to establish alternative maritime trade routes. These expeditions were placed under the command of a Muslim eunuch, Zheng He (Cheng Ho). On his first voyage, Zheng He recognized the Malay state of Melaka as the legitimate successor to China's earlier vassal in the South Seas, Srivijaya, and as the official port for all western vessels seeking to trade with China. The Ming emperors abruptly lost interest in maritime trade again in 1435, but Zheng He's expeditions left behind an enduring legacy in the form of large colonies of Chinese along the straits of Melaka and the north coast of Java. Islam became the state religion in many of these hybrid Chinese-Malay city-states.

The ruler of Melaka converted to Islam in 1436. As the acknowledged successor to the original Austronesian empire of Srivijaya and as the first major coastal power to convert to Islam, Melaka was the most prestigious of the new Muslim sultanates during the fifteenth century. Its enormous ships sailed all over the Indian Ocean, probably reaching as far as Madagascar (Manguin 1993). Melaka's hegemony was abruptly terminated by the Portuguese, whose ships enjoyed a decisive superiority in firepower. Melaka fell to the Portuguese in 1511, who went on to attack the sultanate of Ternate in 1512 and Hormuz, at the mouth of the Persian Gulf, in 1515 (Chaudhuri 1985: 68–69). Between 1510 and 1540 they were able to prevent almost all spices from reaching the Muslim Middle East by blockading the entrance to the Red Sea.

Elite Mysticism in Southeast Asia

This sudden irruption of a non-Muslim power into the Indian Ocean posed a challenge to the Muslim states of the region comparable to that posed by the Mongols in the thirteenth century. And just as the Mongol conquest of Baghdad had caused a diaspora of Arab Muslims to spread throughout the Indian Ocean basin, so the Portuguese conquest of Melaka caused a diaspora of Malay Muslims to spread throughout Island Southeast Asia. This diaspora ultimately contributed to the development of powerful new Islamic sultanates in Aceh, Banten, and, a century later, Gowa.

Like the sultans of Delhi, the early sultans of Aceh turned to the charismatic authority of cosmopolitan *shaikhs* to legitimate their political authority. The most original and influential of these *shaikhs* was Hamzah ibn Abdullah al-Fansuri (d. 1527), a native of Barus in western Sumatra. Hamzah traveled extensively and often mentioned Pasai, Sri Lanka, Mecca, Baghdad, Sinai, and Jerusalem in his writings. He was fluent in Malay, Arabic, and Persian, all languages he could have learned in Barus before he set out on his travels (Guillot and Kalus 2000). Hamzah's teachings were based on Sufi masters ranging from al-Bistami (d. 874) to Ibn al-Arabi (d. 1240), al-Jili (d. ca. 1420) and the Persian Nur al-Din Jami (1414–1492). Jami is the latest author cited in Hamzah's works and Hamzah might even have studied under him. Hamzah was also a contemporary of Shah Ismail Safavi (1484–1524) and could have visited Iran during the period when Ismail was claiming to be an exemplar of Ibn al-Arabi's concept of the Perfect Man.

Hamzah followed al-Jili's interpretation of Ibn al-Arabi's doctrine of the Unity of Being (al-Attas 1970: 69–71; compare Johns 1961: 42). The goal of God's final and most complete creation, humanity, is to reascend the five grades of being in order that it may fully know and love the Creator (Bowen 1987). This can only be accomplished in this life by a mystical adept, *al-Insan al-Kamil*, the Perfect Man. At the beginning of the sixteenth century, it was normally assumed that mystical adepts who claimed the status of the Perfect Man were distinct from and complementary to political rulers. At the end of the century, absolute rulers all over monsoon Asia were claiming this status for themselves. The theological justification of Sultan Iskandar Muda's absolute power in these terms was provided by a follower of Hamzah Fansuri known as Shams al-Din of Pasai (d. 1630). Shams al-Din used same model of the ruler as the Perfect Man to legitimate the autocratic rule of Sultan Iskandar Muda that Abu'l Fazl used to legitimate the autocratic rule of Akbar (Lombard 1967: 158; Reid 1975, 1993).

The political appropriation of the doctrine of the Perfect Man is exemplified in this panegyric written by a court mystic in honor of the sultan of Aceh. This verse was once attributed by al-Attas to Hamzah Fansuri himself, but is now thought to have been written by a later imitator (Braginsky 1999: 143).

Shah 'alam, raja yang adil	World Ruler, the Just King
Raja Qutub yang sampurna kamil	Royal Axis, completely perfect
Wali Allah, sampurna wasil	Friend of God, in complete union
Raja arif, lagi mukammil	Gnostic king, also most Excellent

(My translation of a stanza cited by al-Attas 1970: 12)

Woodward has shown how ideas very similar to those developed in India by Akbar and in Aceh by Iskandar Muda formed the ideological basis of the empire of Mataram, founded by Senapati in Java in around 1584. The *Serat Cabolek* teaches that "the king is the representative of Muhammad, and through him, of Allah," therefore any attack on the *shariah* was viewed as an act of treason (Woodward 1989: 155). But the sultan also reserved to himself the authority to overrule the *ulama* and to decide how the *shariah* should be interpreted. This was a power that few rulers had claimed since the temporal decline of the Abbasid caliphs in the ninth century. Having proceeded further along the mystical path, the sultan was himself above the *shariah* law. But he had to ensure that those lower in the social and religious hierarchy obeyed it. In a series of wars against the *ulama*-dominated coastal cities between 1613 and 1645, Sultan Agung finally succeeded in enforcing this view of the sultan's prerogatives on the empire as a whole.

During the seventeenth century, the five-stage system developed by al-Jili and taken over by Hamzah Fansuri was supplanted by a seven-stage system that became the standard interpretation of Ibn al-Arabi's cosmology throughout South and Southeast Asia. The first extent statement of this system is found in a work written in 1590 by a Gujarati mystic, Muhammad ibn Fadl Allah al-Burhanpuri (1545–1620). It appears in the work of the Sumatran Sufi Shams al-Din at least as early as 1601 (Bowen 1993: 112 n.4). Anthony Johns has argued that while it is usually assumed that Fadl Allah originated the scheme and that Shams al-Din borrowed it, it is just as plausible that the influence went in the other direction (Johns 1965: 9).

According to this sevenfold system, the first three Grades of Being are much the same as those found in the works of al-Jili and Hamzah Fansuri. They refer to the eternal and uncreated grades of Being that exist only in the divine consciousness (Braginsky 1990: 109–110).

1. *Ahadiyyah*: absolute, unmanifested incomprehensible unity.
2. *Wahda*: synthetic unity of Being.
3. *Wahidiyya*: analytical being or unity in multiplicity.

The next four grades involve created beings that are subject to destruction. They are ranked according to their degree of "subtlety" and internal unity:

4. *Alam*: the created world
   a. *Alam al-arwah*: the world of spirits
   b. *Alam al-mithal*: the world of ideas
   c. *Alam al-asham*: the world of physical bodies
   d. *Alam al-insan*: the world of humans

By grouping the created worlds together in this way, Indonesian thought was thus able to reduce a sevenfold scheme into two nested fourfold schemes, allowing them to be mapped onto a wide variety of other fourfold schemes. For example, the law (*shariah*) governs the world (*alam*) and is correlated with water; the mystical path (*tariqa'*) governs analytical being (*wahidiyya*) and is correlated with air, the truth (*haqiqa*) governs the synthetic unity of Being (*wahda*) and is correlated with earth and gnosis (*ma'rifa*) governs the absolute (*ahadiyyah*) and is correlated with fire.

This system of fourfold correspondences formed the basis for popular seventeenth-century allegorical works such as the *Hikayat Shah Mardan*, a text that was translated from Malay into Makassar soon after the conversion of the rulers of Gowa and Tallo' to Islam in 1605 (Braginsky 1990; Arief 1981). Allegorical works such as this made Sufi cosmology accessible on a popular level and the sevenfold system of Shams al-Din is embedded in village rituals throughout Indonesia today.

The Conversion of South Sulawesi

The kingdom of Gowa was founded in the thirteenth century as a small cluster of nine villages on the banks of the Jene'berang river in South Sulawesi. By the beginning of the sixteenth century, Makassar ships were sailing as far as the Malay Peninsula. When the Portuguese conquered Melaka in 1511, the kings of Gowa and Tallo' welcomed Muslim refugees, but they were also happy to establish diplomatic and trade relations with the Portuguese. They used Portuguese guns to centralize their power over South Sulawesi in an unprecedented way. Gowa no longer carried out war merely to seize prestige trade goods such as gold and jewels, but to capture slaves that could be put to work on large-scale irrigation and fortification projects (see Gibson 2005: 152–156). By 1560 Gowa had reduced almost the whole of South Sulawesi to tributary status. The major exception was the Bugis kingdom of Bone on the eastern side of the peninsula (see figure 2.1).

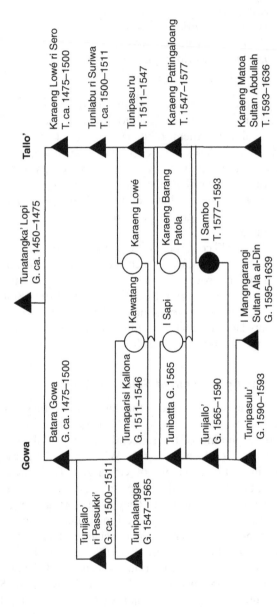

Figure 2.1 The Kings of Gowa and Tallo', 1450–1639

Gowa

Tunatangka' Lopi
G. ca. 1450–1475

Batara Gowa
G. ca. 1475–1500

Tunijallo'
ri Passukki'
G. ca. 1500–1511

Tumaparisi Kallona
G. 1511–1546

I Kawatang

Karaeng Lowé

Tunipaalangga
G. 1547–1565

Tunibatta G. 1565

I Sapi

Karaeng Barang
Patola

Tunijallo'
G. 1565–1590

I Sambo
T. 1577–1593

Tunipasulu'
G. 1590–1593

I Mangngarangi
Sultan Ala al-Din
G. 1595–1639

Tallo'

Karaeng Lowé ri Sero
T. ca. 1475–1500

Tunilabu ri Suriwa
T. ca. 1500–1511

Tunipasu'ru
T. 1511–1547

Karaeng Pattingaloang
T. 1547–1577

Karaeng Matoa
Sultan Abdullah
T. 1593–1636

By the 1590s, Gowa was administered by a relatively sophisticated bureaucracy. At the top was the *Somba* or Emperor of Gowa, assisted by a *Tuma'bicara Butta* or Chancellor, who was usually the king of Tallo'. Under them were two *Tumailalang*, or Ministers of the Interior, with the rank of *Karaeng*. Beneath the *Tumailalang* was the *Sabarana* (from Persian Shahbandar), Harbor Master, and the *Tumakkajananngang*, Guild Master. They mediated between the throne and the nine members of the *Hadat*, or Council of Electors, also known as the *Bate Salapang*, the Nine Banners. These electors were the chiefs of the villages that formed the core of the original kingdom of Gowa. They held the lesser rank of *gallarrang*.

In 1593, the royal chancellor of Gowa, I Malinkaeng, deposed the ruler of Gowa and replaced him with his seven-year-old brother, I Mangngarangi (r. 1593–1639). I Malinkaeng then deposed his own sister, I Sambo, and had himself installed as king of Tallo'. Twelve years later, it was I Malinkaeng who led the drive to convert the whole of South Sulawesi to Islam.

Muslim traders were probably already present in South Sulawesi during the fifteenth century, but conversion to Islam by the rulers of the main Makassar and Bugis states did not occur until 1605–1611. Throughout the sixteenth century, the rulers of South Sulawesi tried to play the Portuguese and Spanish off against the Muslim rulers of Southeast Asia. The rulers of South Sulawesi held out against both Islam and Christianity until February, 1605, when the king of the Bugis state of Luwu' was converted to Islam by a Sumatra *shaikh* called Suleiman (Noorduyn 1956). Christian Pelras suggests that this conversion may have been facilitated by the way Suleiman integrated Islamic cosmology with the myths contained in the great epic of the Bugis people, the *I La Galigo*, identifying Allah with the Bugis supreme being Dewata Seuwae. Suleiman may have brought with him from Sumatra Hamzah Fansuri's idea that the doctrine of the Unicity of God (*Tauhid*) could be interpreted to mean that local religious traditions like that of the Bugis were attempting to approach the same ultimate reality by different paths (Pelras 1985: 120).

The following myth of the subsequent conversion of Gowa and Tallo' was published by Matthes in 1885.

The Conversion of I Malinkaeng

On September 22 a great ship was seen approaching Tallo'. But as it neared, it appeared to be only a brig, and then it shrank in size to an ocean-going sloop to a two-masted cargo ship, to a small fishing boat, and finally to a map of the world. On it was seated a Minangkabau from Kota-Tanga called Dato-ri-Bandang. As soon as he came ashore, he began to pray in the Muslim manner, making use of a rosary, and to recite the Koran. The people of Tallo' informed the king, I Malinkaeng. The latter set off at dusk to see the visitor for himself. As he entered the port at the Fort of Tallo', he saw five men standing together. The middle one was standing on a flat white stone, and asked the King in an imposing tone where he was going. When the ruler told him he was going to see Dato ri Bandang, the man told him to give him his greetings and said that he was the Prophet Muhammad. The king had never yet heard of Muhammad, and had difficulty understanding his foreign name. To help him remember, the Prophet wrote on his hands the Arabic words, "*bismillahir-rahmani r rahimi*," i.e. in the name of Allah the All-Merciful, as well as "*assalamu alaika, wa-rachmatu-l-lahi wa-barakatuhu*," i.e. "Peace be upon you, Allah's mercy and blessing be upon you!" The Prophet used neither ink nor pen, but only the spittle from his mouth on his right index finger to write these words. And yet the light of the letters was brighter than a full moon. And a wonderful odor, which far surpassed all the perfume and incense of Sulawesi, spread through the whole area. Then the five exalted ones disappeared and the King of Tallo' continued on his way.

When the king reached Dato-ri-Bandang, he asked him what Supreme Being he worshipped. "My God is your God," answered Dato-ri-Bandang. "He is the Lord of the seven levels of the Heavens and the Earth. He has created everything that there is. He is the Lord of Life and Death, Who was from the beginning and Who never had a beginning. He will also be the last of all that exists, and there is no one who shall be after Him."

The King immediately requested instruction. It required only a few hours for him to learn not only the profession of faith of the Muslim, that there is no God but Allah and that Muhammad is his Prophet, but also the Muslim manner of prayer with all the accompanying gestures, as well as the proper recitation of the Koran. When the instruction was completed, Datu-ri-Bandang placed one hand upon the head of the king, and the other under his chin, and turned his gaze up to Heaven. And when Dato-ri-Bandang asked him what he now saw, he answered: "I see the throne of Allah, as well as the table *lou-l-mahapul*, on which the deeds of men both good and evil are noted down. And Allah asks of me that I embrace Islam, and also bring the others to it, and wage war on those who oppose me in this." Thereupon Datu-ri-Bandang who still held the head of the ruler fast turned his gaze downwards and asked him again what he now saw. "I see," said the ruler, "to the furthest depths of the Earth and there I see Hell, in which Allah wills that I and others shall be placed if they show themselves reluctant to accept your teaching."

The king had thus far forgotten to deliver the greetings of the great Prophet of Mecca to Dato-ri-Bandang. But now he suddenly remembered to do so. When Dato-ri-Bandang observed the radiance from the King's hands, he realized the King was far above him, in spite of his own high mission, since he had himself never received an appearance from the Prophet. Soon all Tallo' and Gowa had embraced Islam, which also speedily spread from there still further through Celebes. The grave of this first apostle of Islam in South Celebes, which is to be found in Kaluku-Bodowa in Tallo', is still held in veneration among the descendants. (condensed from Matthes [1885] 1943: 387–390)

Mattulada noted in 1976 that the place where Muhammad appeared to I Malinkaeng continues to be regarded as a place of power (*karama'*) where it is beneficial to perform acts of veneration (*siara*) (Mattulada 1976: 14).

This myth sets up a parallel between the traditional authority I Malinkaeng derived from his connection to a local *Tomanurung*, and the charismatic authority he acquired from his direct contact with the universal prophet, Muhammad. Muhammad appears to him standing on a flat white stone, exactly like the *Tomanurung* who are encountered at the beginning of the foundation myths of most local dynasties in South Sulawesi (Gibson 2005). The Chronicle of Tallo' makes a similar point about the dual basis of I Malinkaeng's authority when it notes that, "he was the first king to be installed with the Al-Quran which was held beside a *kalompoang* (regalia) of Gowa, the Sudanga." The Sudanga was the sword left to the kings of Gowa by Lakipadada, brother-in-law of the divine princess who founded the kingdom (Manyambeang and Mone 1979: 17; Wolhoff and Abdurrahim [1960]: paragraph 4).

Muhammad transferred his charisma by using his saliva to inscribe the fundamental precept of Islam, the *fatiha*, on the body of I Malinkaeng. What is for ordinary people an impure bodily by-product is for *shaikhs* and prophets supernaturally pure, fragrant, and radiant. In Lambek's phrase, "objective" Islamic knowledge was directly "embodied" in I Malinkaeng, the transcendental word of God was made immanent (Lambek 1993: 149–155).

Matthes noted that the four companions of the prophet must have been none other than the first four caliphs, Abu Bakr, Usman, Umar, and Ali. The apparition thus served not only to convey the blessings of the Prophet directly on the Karaeng of Tallo', but also to mark him as another in the line of Gods' representatives on earth, or caliphs. It is clear that the story intends to place the pious ruler above any *wali Allah*, "friend of God." The model of royal power is thus close to that claimed at the time by Akbar in India or by Agung in Java.

Muhammad's blessing allows I Malinkaeng to undergo a greatly accelerated course of instruction at the hands of the foreign emissary, Datu ri Bandang. He is first instructed in the outer ritual forms of the faith: the profession of faith, the daily prayers, and the recitation of the Koran. These are all part of the religious law, *shariah*. The Chronicle of Tallo' stresses I Malinkaeng's mastery of the outer forms of Islam. "He was literate in Arabic, performed his both mandatory and supererogatory prayers regularly, refused alcohol, paid the religious tax on his holdings in gold, buffalos and rice and studied many works on Arabic morphology with Khatib Intang and Manawar the Indian" (Noorduyn 1987a: 315).

In the conversion myth, I Malinkaeng proceeds immediately from the outer path of the *shariah* to the inner path of mystical insight, *tariqa*. He gazes upon the Throne of Allah in the heavens and upon the depths of Hell in the earth. The visiting *shaikh* ends by acknowledging the superior status of the ruler when he learns of the latter's direct encounter with the Prophet.

The myth of the conversion of I Malinkaeng follows fairly closely in these respects the myths of the conversion of the ruler of Melaka and of the ruler of Pasai contained in the *Sejarah Melayu* and in the *Hikayat Raja Raja Pasai*, respectively (Jones 1979). According to Anthony Johns, the oldest surviving manuscript of the *Sejarah Melayu* dates to about 1612, and the *Hikayat Raja Raja Pasai* was actually written later. If this is correct, it means that these texts actually give us an idea of Islam as it was understood in Sumatra just at the time Sultan Abdullah of Tallo' converted (Johns 1975: 40).

In the *Sejarah Melayu*, Raja Tengah sees Muhammad in a dream. Muhammad teaches him the profession of faith, gives him the name Muhammad, and tells him to welcome the man who will arrive from Jeddah on a ship the next day. This was Sayyid Abd al-Aziz, who stayed on to instruct the king, now called Sultan Muhammad Shah, in all the details of Islam.

In *Hikayat Raja Raja Pasai*, Merah Silau has a dream in which Muhammad appears to him, spits in his mouth, and gives him the name Malik al-Saleh. Immediately he is able to recite the profession of faith and the entire Koran, and is magically circumcised. Forty days later a ship arrives carrying Shaikh Ismail from Mecca, regalia sent by the caliph, and a *fakir*, an ascetic mystic, from Mengiri in India. Shaikh Ismail is sent back to Mecca with gifts for the caliph, and the *fakir* stays behind to implement the Islamic law in full (Jones 1979: 133–136).

Like the Tallo' myth, the Pasai myth includes a reference to Muhammad's saliva. In this case, the saliva is used to transmit oral as opposed to textual knowledge. Both the *Hikayat Raja-Raja Pasai* and the *Hikayat Aceh* begin with the myth of origin of the royal dynasty from a princess who emerges from a bamboo (see Gibson 2005: 125–141). As with the dynasties of Tallo' and Gowa, there is no attempt in the royal chronicles to deny the indigenous supernatural origins of the ruling line or to manufacture genealogies linking rulers to the Islamic Middle East. By adopting the Sufi idea of Muhammad's direct manifestation to the spiritual elect, these dynasties were able to claim recognition from the cosmopolitan Islamic community without abandoning their claim to originate from local supernatural beings. They are also able to treat instruction by foreign *ulama* in the external details of religion as secondary importance in relation to the direct . transmission of the inner truths of religion by the Prophet himself.

The myth of I Malinkaeng's conversion to Islam illustrates the manner in which the ruling dynasty was able to make a certain version of Islamic doctrine compatible with the indigenous cult of the divine royal ancestors that had given it legitimacy up to that point. King I Malinkaeng of Tallo' may be taken as a paradigmatic example of an authority figure who combined the traditional authority conferred by descent from the royal ancestors with the charismatic authority conferred by the personal instruction he received from the Prophet Muhammad. He passed both forms of authority on to his successors, who continued to claim descent from the Tomanurung and a status equal to that of the early caliphs. From 1605 on, the kings of South Sulawesi increasingly relied on a range of Islamic scriptures, charismatic *shaikhs*, and mystical practices, both to legitimate themselves in the eyes of their subjects and to forge new alliances with other Islamic rulers throughout Island Southeast Asia.

The Imperial Expansion of Gowa, 1605–1667

By the beginning of the seventeenth century, I Malinkaeng was ready to transform Gowa-Tallo' from an agrarian into a maritime empire. But to do so he needed to acquire a new sort of authority over his own subjects, and a new sort of legitimacy in the commonwealth of Indonesian kingdoms. This he accomplished by converting to Islam. He quickly persuaded his cousin and coruler, the Karaeng of Gowa I Mangngarangi, to do so as well. Two years later, in 1607, the entire *Hadat*, Royal Council, of Gowa converted and the empire officially became a sultanate. After his conversion, I Malinkaeng added the title of Sultan Abdullah Awwal al-Islam, or "Slave of God, the First in Islam" to his other titles. He became best known, however, as Karaeng Matoaya, the Old Lord, in contrast to his coruler, Karaeng I Mangngarangi of Gowa, who was thirteen years his junior. I Mangngarangi took the title of Sultan Ala al-Din of Gowa. As rulers of the mightiest state in South Sulawesi, it fell to these men to persuade all other rulers on the peninsula to convert to Islam, either voluntarily or by force (see map 2.1). The one exception was the Datu of Luwu', who was acknowledged as the only ruler to have converted before Sultan Abdullah.

In 1608 Sultan Abdullah invited the ruler of Bone to convert, but was rebuffed. He launched two unsuccessful attacks against the "Triple Alliance" of Bone, Wajo', and Soppeng. Finally, in alliance with Luwu', Gowa forced Sidenreng to convert in 1609, Soppeng in 1609, Wajo' in 1610, and Bone in 1611. Within six years of his own conversion, Sultan Abdullah of Tallo' was acknowledged by the whole of South Sulawesi as *Awwal al-Islam*, the "First in Islam."

The Islamic wars through which Sultan Abdullah forced the submission of all rulers to Islam did not lead to an immediate increase in the centralization of political or economic power on the peninsula. According to the Chronicle of Tallo', Abdullah was especially careful not to link forced conversion to Islam with increased tribute payment or with a loss of political autonomy (Manyambeang and Mone 1979: 16). Outside the core region of Gowa and Marusu', agricultural production continued to be decentralized, and the sources of legitimacy of local dynasties continued to derive from highly localized Tomanurung myths.

What was new was the global legitimacy conferred on Sultan Abdullah as *Awwal al-Islam*, "First in Islam." David Bulbeck remarks that Sultan Abdullah may have been the highest-ranking Makassar ever to live. His chief wife was the only one ever to receive an honorific nickname, a usage otherwise reserved only for the greatest pre-Islamic rulers of Makassar, Tallo', Marusu', and Sanrabone (Bulbeck 1992: 44). So great was the prestige of Tallo' during his lifetime that the hypergamous flow of women from the royal house of Tallo' to that of Gowa ceased. Between the defeat of Tallo' by Gowa in 1535 and the exile of Tunipasulu' in 1593, numerous princesses of Tallo' had married the kings of Gowa, but only one princess of Gowa had married a king of Tallo'. Between 1593 and Sultan Abdullah's death in 1636, no princess of Tallo' married a king of Gowa (Bulbeck 1992: 129). While the rulers of Gowa and Luwu' also took the title of Sultan upon their conversion, the prestige of the royal house of Tallo' exceeded that of

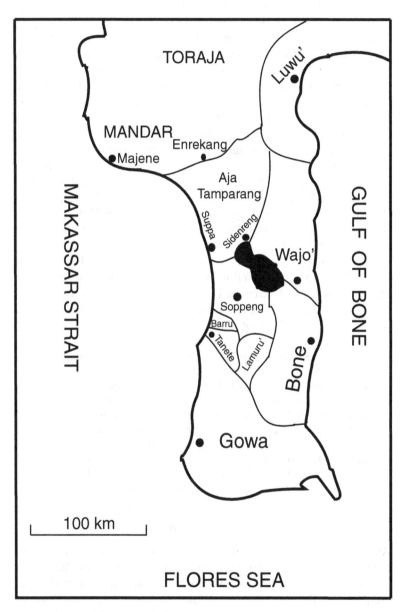

Map 2.1 South Sulawesi Kingdoms, 1605

Gowa and Luwu' for the first time. The rulers of Soppeng and Bone assumed the title of Sultan only after the defeat of Gowa-Tallo' by the VOC in the 1660s (Bulbeck 1992: 44).

The real increase in Sultan Abdullah's political power came not from a further centralization of power over the landmass of South Sulawesi, but from a projection of Gowa's power overseas. After the final submission of the Bugis kingdoms to Islam in 1611, he turned his attention to maritime expansion. Beginning with an attack on Bima in 1616, his navies went on to dominate the entire eastern archipelago (Manyambeang and Mone 1979: 17).

These maritime conquests required a set of innovations in naval technology made available by a fusion of indigenous and European boat-building techniques, which occurred first in the Spanish shipyards of Cebu and Ternate. Bira, with its long history of colonization from the Moluccas and its strategic location on the Java-Moluccan trade routes is a good candidate for having introduced these innovations (Horridge 1979: 51).

The sultans of Gowa maintained their claim to the status of Perfect Man until their defeat by the VOC in 1667. In the *Sya'ir Perang Makassar*, an account of Sultan Hasan al-Din's downfall in 1667, his Malay scribe, Ince Amin, described him in the same terms used to depict the sultan of Aceh.

> My lords, hear a humble homage to the most magnificent king; perfect in gnostic understanding [*'arif*], caliph of the annihilators of being [*fana*]. By the grace of God and the intercession of the Prophet, caliph of God in the two states; beloved by God and His friends [*wali*], there was joy and wealth in both realms. World ruler and just king, royal caliph whose perfection is complete; friend of God whose communion is total, both gnostic master and without fault. By the grace of God, Creator of the world, Who raised up the two worlds; Whose numerous community fills the world with joy and prosperity day and night. (My translation of stanzas 144–148 in Skinner 1963: 110–112)

Conclusion

At the end of the sixteenth century, Austronesian kings all over Southeast Asia sought charismatic authority by portraying themselves as exemplars of the Perfect Man while retaining their traditional authority as the descendents of local divinities and custodians of sacred heirlooms. Between 1605 and 1611, the kings Tallo' and Gowa imposed a hierarchical and mystical form of Islam on all the kingdoms of South Sulawesi. During the next half-century, they expanded their realm from an inland kingdom to a maritime empire that encompassed all of eastern Indonesia. A similar synthesis of charismatic and traditional sources of power and authority occurred at the village level as ordinary peasants converted certain sacred sites into the tombs of Islamic *shaikhs* while preserving others as shrines to the local royal ancestors.

In most of South and Southeast Asia, the ideology of the ruler as Perfect Man eroded during the seventeenth and eighteenth centuries as the British and Dutch East India Companies humbled one sultan after another. The religious supremacy of the king survived somewhat longer in central Java than it did elsewhere, but only because the Dutch acquired such complete control of the seas that Java was cut off from new developments in the wider Islamic world. The first signs of these developments arrived in South Sulawesi during the 1640s.

Chapter 3

Cosmopolitan Islam in South Sulawesi, 1640–1705

In the1980s Abdul Hakim provided me with copies of several old manuscripts that turned out to be Makassar translations of writings by Nur al-Din al-Raniri (Nuruddin ar-Raniri [1642] 1983). Nur al-Din al-Raniri was a scholar of Hadrami descent who was born in Gujarat, probably to a Malay mother. He may have been trained from birth to serve as a missionary in the East Indies. He was the author of the most extensive body of Islamic writings ever produced in the Malay language. He made it his business to replace the pantheistic teachings of Hamzah Fansuri in the royal courts of Southeast Asia with the synthesis of *tariqa'* and *hadith* studies that had developed in Mecca and Medina during the sixteenth century.

A manuscript I obtained in Bira in 2000 contained two *silsilas*. One showed that al-Raniri had initiated Haji Ahmad al-Bugisi into the Qadiri Sufi order in the mid-seventeenth century. Haji Ahmad's student, Abd al-Rahman of Lamatti, settled in Bira. The other *silsila* showed that one of al-Raniri's teachers, Ibrahim al-Kurani, had initiated Ibrahim Barat of the Bugis kingdom of Bulo Bulo into the Shattari Sufi order at about the same time. Ibrahim Barat's student, also called Abd al-Rahman, settled in Selayar. These *silsilas* indicate that the model of Islam al-Raniri taught in Aceh between 1637 and 1644 arrived in Bira and Selayar within a few decades of the original conversion of the kings of South Sulawesi.

To understand the origins of al-Raniri's form of Islam, we must turn to the cosmopolitan world of mercantile city-states that lined the coasts of the Arabian Sea in the sixteenth and seventeenth centuries. This world was quite different from that of the great inland empires of Iran and India at the time. Rulers depended on the goodwill of long-distance traders who moved in and out of their ports. They derived both their economic and religious authority from their reputation for enforcing the universal norms of the *shariah* law as interpreted by a cosmopolitan group of *ulama*. The model of kingship taught by the *ulama* assigned a relatively humble role to local political rulers. They were expected to establish their religious legitimacy not by claiming superior legal and mystical knowledge for themselves, but by patronizing charismatic scholars and mystics who traveled far and wide through the Islamic world accumulating universal knowledge.

The first generation of pilgrims from South Sulawesi encountered this synthesis all along the trade routes to and from the Hejaz during the early seventeenth century. The *hajjis* returned home with a very different model of the ruler's authority than the one being promoted in the royal courts, a development that had immediate political repercussions. In his monumental study of "Networks of Malay-Indonesian and Middle Eastern *Ulama* in the Seventeenth and Eighteenth Centuries," Azyumardi Azra argues that there was an influx of Indian Sufis into Mecca and Medina in the seventeenth century, and that their mystical teachings fused with the legal teachings of north African and Egyptian experts in *hadith* studies (Azra 2004: 31). Azra borrows the term "neo-Sufism" from Fazlur Rahman to describe this fusion of *tariqa'* and *hadith* studies. Rahman introduced "neo-Sufism" to refer to an "orthodox" reform of Sufism that began in the eighteenth century in which mysticism "was stripped of its ecstatic and metaphysical character" and acquired an outward-looking and reformist orientation (Rahman 1979: 206). He traced the origin of neo-Sufism back to the teachings of Ibn Taymiya (d. 1328) in the fourteenth century (Rahman 1979: 195).

Sean O'Fahey and Bernd Radtke question the notion that there was a neo-Sufi movement in the eighteenth century, and that the writings of the central Sufis of that period show any significant discontinuity with those of previous generations. They deny that even the most innovative of the eighteenth-century African "neo-Sufis," Ahmad ibn Idris (d. 1837) and Ahmad al-Tijani (d. 1815), embodied "the neo-Sufi cliché's postulate of innovative discontinuities with the Sufi past, the rejection of the mystical philosophical tradition of Ibn al-Arabi, of the initiatory path, of the chains of spiritual authority, all in favour of some kind of 'Sufi Wahhabism' " (O'Fahey and Radtke 1993: 54–55; compare Trimingham 1971: 106). They go on to argue that the notion that there was a fundamentally new kind of Sufism in the eighteenth century is an illusion generated by European colonialists who saw the Sufi orders as one of the principal sources of resistance to

European rule. In fact, the Sufi orders continued to follow much the same doctrines and practices that they had in the past. Their new "activism" was simply a response to foreign intervention.

In response to O'Fahey and Radtke, I would argue that there were significant innovations in Sufi thought, but that they occurred far earlier than Rahman thought, in the sixteenth-century encounter between Indian *shaikhs* and North African *ulama*. During the sixteenth century, a wave of Indian Sufis who objected to the doctrines of mystical kingship that were being developed in Akbar's court sought refuge in Mecca and Medina. There they encountered the primarily Middle Eastern *ulama* who were teaching the four orthodox schools of Islamic law. A new synthesis of legal and mystical studies developed in this environment. Students came to pay equal attention to learning *hadith* and to mystical practice. Simultaneous immersion in the ritual techniques of a Sufi *tariqa'* and in the *hadith* tended to reorient the imaginative life of the student away from the founder of his particular mystical order and toward the example of the Prophet Muhammad, the founder of the *sunna*, the path every human should follow to achieve salvation.

Muslims from Kurdistan, Southern Arabia, Western India, East Africa, and Southeast Asia tended to follow the school of law founded by al-Shafii (767–820) and were thrown together when they studied in the Hejaz. From this time on, most legal scholars also became mystical masters, a pattern that continues to hold true for most Javanese *kyai* (Lukens-Bull 2005). They tended to join several Sufi orders, among the most important of which were the Naqshbandiyya, the Qadiriyya, the Shattariyya, and the Khalwatiyya. These crosscutting legal and mystical affiliations created an intricate network of personal relationships that spanned the Indian Ocean. Figure 3.1 shows the principal *shaikhs* and rulers I discuss in this chapter.

The religious authority of these networks was independent of the authority of any particular political ruler. This made the networks particularly well suited to serve as a source of resistance to the imposition of political regimes by rulers who had little claim to legitimacy in terms of either traditional or charismatic authority. This was the case whenever and wherever European colonial regimes first appeared on the scene. This happened first along the coasts of the Indian Ocean when Portuguese warships disrupted traditional trading patterns at the beginning of the sixteenth century. In Island Southeast Asia, the attempt by the VOC to impose a monopoly over maritime trade was well underway by the 1630s. In North Africa, colonial intervention did not begin until the eighteenth century.

58

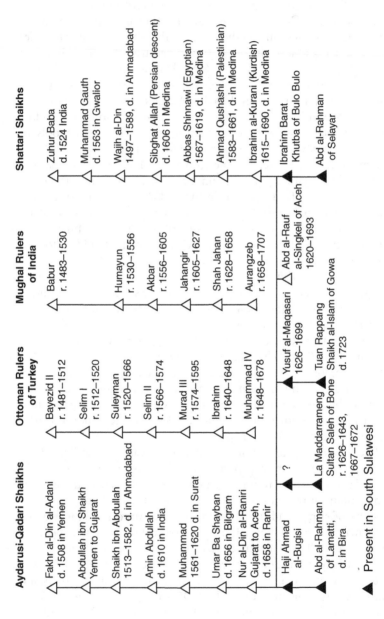

Figure 3.1 Spiritual and Dynastic Genealogies, 1500–1700

◀ Present in South Sulawesi

Aydarusi-Qadari Shaikhs

◁ Fakhr al-Din al-Adani
d. 1508 in Yemen

◁ Abdullah ibn Shaikh
Yemen to Gujarat

◁ Shaikh ibn Abdullah
1513–1582, d. in Ahmadabad

◁ Amin Abdullah
d. 1610 in India

◁ Muhammad
1561–1620 d. in Surat

◁ Umar Ba Shayban
d. 1656 in Bilgram

◁ Nur al-Din al-Raniri
Gujarat to Aceh,
d. 1658 in Ranir

◀ Haji Ahmad
al-Bugisi

◀ Abd al-Rahman
of Lamatti,
d. in Bira

Ottoman Rulers of Turkey

◁ Bayezid II
r. 1481–1512

◁ Selim I
r. 1512–1520

◁ Suleyman
r. 1520–1566

◁ Selim II
r. 1566–1574

◁ Murad III
r. 1574–1595

◁ Ibrahim
r. 1640–1648

◁ Muhammad IV
r. 1648–1678

◀ Yusuf al-Maqasari
1626–1699

◀ Tuan Rappang
Shaikh al-Islam of Gowa
d. 1723

? ◀
◀ La Maddarrameng
Sultan Saleh of Bone
r. 1626–1643,
1667–1672

Mughal Rulers of India

◁ Babur
r. 1483–1530

◁ Humayun
r. 1530–1556

◁ Akbar
r. 1556–1605

◁ Jahangir
r. 1605–1627

◁ Shah Jahan
r. 1628–1658

◁ Aurangzeb
r. 1658–1707

◁ Abd al-Rauf
al-Singkeli of Aceh
1620–1693

Shattari Shaikhs

◁ Zuhur Baba
d. 1524 India

◁ Muhammad Gauth
d. 1563 in Gwalior

◁ Wajih al-Din
1497–1589, d. in Ahmadabad

◁ Sibghat Allah (Persian descent)
d. 1606 in Medina

◁ Abbas Shinnawi (Egyptian)
1567–1619, d. in Medina

◁ Ahmad Qushashi (Palestinian)
1583–1661, d. in Medina

◁ Ibrahim al-Kurani (Kurdish)
1615–1690, d. in Medina

◀ Ibrahim Barat
Khutba of Bulo Bulo

◀ Abd al-Rahman
of Selayar

In much of the African interior, it did not take place until well into the nineteenth century.

The doctrines and networks created by these scholarly mystics thus acquired a political dimension in different times and places depending on when and where European colonialism appeared. The association between "neo-Sufism" and anticolonial resistance is a by-product of the autonomy the cosmopolitan networks enjoyed from local political systems. But political resistance was neither the source nor the goal of the networks, which remained centered on the acquisition and propagation of religious knowledge whether the local political order was viewed as legitimate or not.

The Teachings of Al-Raniri

Nur al-Din was born in Rander, a village just outside Surat in Gujarat, to a Hadrami father and to a mother who may have been Malay. As was the norm in this milieu, he probably received his early education back in the Hadramaut. We know that he performed the *hajj* in 1620/1621 and sailed for Southeast Asia not long afterward (al-Attas 1986: 7). When Sultan Iskandar Thani II came to power in 1637, he appointed al-Raniri his *Shaikh al-Islam*, the highest religious office in the realm. He commissioned him to write an encyclopedic work in Malay on universal history, the *Bustan al-Salatin*. According to al-Attas, it is the longest work ever written in Malay (al-Attas 1966, 1986).

Between 1638 and 1641 al-Raniri wrote "The Proof of the Truthful in the Refutation of the Heretics," a polemical attack on the works of Hamzah Fansuri and Shams al-Din. With the backing of the sultan, he burned their books and persecuted their followers to cleanse the realm of what he viewed as a pantheistic interpretation of the *wahdat al-wujud*. Not surprisingly, he made a number of enemies among the local Islamic establishment.

Sultan Iskandar Thani died in February, 1641 and was succeeded by his wife. At first, al-Raniri retained his influence over the sultana. In 1642 he wrote one of his most influential works, *Akbar al-Akhirah fi Ahwal al-Qiyamah*, "The Afterlife and the Day of Judgment." It was translated from Malay into many other Indonesian languages, including Makassar (al-Attas 1986: 27). This is the text that acquired a central place in funeral rituals in Ara and Bira.

On August 8, 1643, a Minangkabau scholar named Saif al-Riyal returned from his studies in Surat. He had been the student of an Acehnese scholar, Shaikh Jamal al-Din, who was executed at the behest of al-Raniri, and had perhaps gone into exile as a result.

Supported by the sultana's consort, al-Raniri denounced Saif al-Riyal as a heretic. The latter gained the backing of the Maharajalela, who presided over the Royal Council. By August 27, 1643, Saif al-Riyal had won the contest (Ito 1978: 491).

Al-Raniri left Aceh soon afterward, returning to Gujarat with his major work on Sufism unfinished, the *Jawahir al-Ulam fi Kashf al Ma'lum*. He died in Rander in 1658. His nemesis Saif al-Riyal did not last long afterward as the *Shaikh al-Islam* in Aceh. He was replaced by one of al-Raniri's Sumatran students in 1661, Abd al-Rauf al-Singkili. Abd al-Rauf left Aceh at the same time as al-Raniri lost favor in 1643. He spent eighteen years in Mecca as the student of Ahmad Qushashi and returned to Aceh upon the latter's death in 1661. It was only after Abd al-Rauf became the leading religious figure in Aceh that al-Raniri's *Jawahir al-Ulam* was completed by one of his students in 1665 (al-Attas 1986: 16–21). Abd al-Rauf's somewhat more tolerant version of al-Raniri's teachings served as the official creed of Aceh throughout the reigns of the four Queens who ruled until his death in 1690.

Cosmopolitan Islam in Bone and the Fall of Gowa, 1640–1696

According to the Qadiri and Shattari *silsilas* I obtained in Bira in 2000, al-Raniri's teachings were brought to South Sulawesi in the mid-seventeenth century (see figures 3.2 and 3.3). The *silsilas* from Bira can be compared with other published *silsilas* to establish their accuracy. The most useful are the *silsilas* left by Shaikh Yusuf (b. 1626, d. 1699), who was initiated into the same branches of the Shattariyya and Qadiriyya as Ibrahim Barat and Hajji Ahmad the Bugis, respectively (Tudjimah 1987: 200, 203; Abu Hamid 1994: 360–363). Another relevant *silsila* is that of Nur al-Din al-Raniri, who appears as the Qadiriyya master of both Shaikh Yusuf and Hajji Ahmad. Al-Raniri belonged to both the Qadiriyyah and Rifaiyya orders. Where the two orders have parallel chains of transmission, his *silsila* overlaps with those of Shaikh Yusuf and Haji Ahmad (al-Attas 1986: 14–15). Al-Attas also gives a separate Qadiriyya silsila for Shaikh Abd al-Qadir al-Malabari, a Sufi he knew in Johore. This man was initiated in 1946 by a pupil of the famous Minangkabau Ahmad Khatib al-Sambasi (1855–1916) who rose to the post of Imam of the Shafii legal school in main mosque at Mecca in 1889 (al-Attas 1963: 54; Noer 1973: 31–33; Laffan 2003). Finally, the works of Rizvi contain an immense amount of detail about these Sufis who traversed the Indian Ocean in the sixteenth and seventeenth centuries (Rizvi 1965, 1978, 1983).

1. Husein b. Ali b. Abu Talib [d. 680 at Karbala]
2. Imam Ali Zain al-Abidin [d. 712]
3. Imam Muhammad b. Ali al-Baqir [d. 731, 5th Shia Imam]
4. Imam Ja'far b. Muhammad al-Sadiq [b. 700, d. 763/5, 6th Shia Imam, prime master of both twelver and Ismaili Shia]
5. Sultan al-Arif Abi Zaidah Sathari
6. Maulana Muhammad al-Arabi
7. Maulana al-Araba
8. Ibnul Mudhaffar
9. Abi al-Hasan al-Arakan
10. Ahad al-Qalbi wa Mawari al-Nahr
11. Muhammad al-Shaq Shattari
12. Muhammad Arif Shattari [of Bukhara]
13. Al-Shaikh Qadi Shattari [Muhammad 'Ala "Qazin" of Bengal]
14. Abu al-Fatah Hidayatullah Sarmillah [died in Patna]
15. Walwar al-Shaikh Dzahra al-Hajj Hasur Hamid
16. Al-Sultan al-Muhakikin Gaus al-Ilmi al-Shaikh Muhammad Hajji al-Hamid [d. 1563, cf. Rizvi: Shaikh Abu'l-Mu'yyad Muhammad ibn Khatir al-Din al-Ghaus al-Hindi]
17. Maulana Wajih al-Din Al-Alubi [al-Ahmadabadi al-Gujarati, d. 1589]
18. Said Sibghat Allah b. Said Ruhullah [al-Barochi al-Madani d. 1606]
19. Saidin b. al-Mawambi Abdullah [cf. Rizvi: Ahmad b. Ali. b. Abd al-Qudus b. Muhammad Abbas Shinnawi, d. 1619]
20. Al-Shaikh Ahmad [Qushashi] b. Muhammad [b. Yunus] al-Madina al-Anshar al-Shaikh bil Qasiah [b. 1583, d. 1661]
21. Shaikh Burhan al-Din [Ibrahim] b. Hasan al-Kurani [b. 1616, d. 1689]
22. Al Shaikh Ibrahim Barat al-Haq Khutba Bulo-Bulo
23. Abd al-Rahman Abdullah Selayar

[These last nine names are difficult to interpret.]

24. Hafza (a woman)
25. Al-Hae yakni Hajari
26. Al-Shaikh Akbari
27. Al-Shaikh al-Gauta
28. Al-Shaikh al-Gauta al-Adimi
29. Al-Shaikh Shaid b. Arabi
30. Al-Shaikh Kadir Sabli
31. [blank]
32. Al-Shaikh Abu al-Gaib

Figure 3.2   The Shattari *Silsila* of Bira

According to the Qadiri *silsila* from Bira, al-Raniri's local student was Haji Ahmad, and he had the following immediate successors:

Q32. Haji Al-Shaikh al-Julaij Ahmad ibn Abdullah al-Bugisiya (*Panre Lohe*, The Greater Authority)

Q33. Al-Shaikh Abd al-Rahman ibn Abdullah Lamatti (*Panre Keke*, The Lesser Authority) [In a marginal note: He who initiated religion in Bira. His tomb is in the yard of Makotta's house in Bira Lohe]

Q34. Shaikh Abd al-Jalil ibn Abdullah Bulo-Bulo (*Guru Toaya*, The Senior Teacher)

1. Ihda Atthabi' in Hasan al-Basri [b. 643, d. 728]
2. Habibullah Hasanji [al-Ajami] [d. 737]
3. Dawud al-Ta'i [d. 781]
4. Ma'ruf al-Karkhi [d. 815]
5. Sari al-Saqati Bigadadi [d. 865/867]
6. Abi al-Qasim Al-Sultan al-Salifihi Junaid al-Bagdadi [d. 910]
7. Al-Istada Abu Bakr al-Shibli [d. 945]
8. Abil Fadli Abd al-Wahid b. Abd al-Aziz al-Thami
9. Abil Fajri Muhammad Abi Abdullah al-Fawzi
10. Abil Hasan Ali [b. Ahmad b. Yusuf al-Ursh al-Hakkari]
11. [Kabir al-Nuhban] Abi Said Mubarak b. Ali al-Maharami
12. Wal-Gauta al-Gauta Muhyi al-Din Shaikh Abd al-Qadir al-Arief Bigadadi wal Jailani [b. 1077, d. 1166]
13. Abdullah b. Ali b. Husna al-Assadi Bigadadi
14. Abihissamit Abdullah b. Yusuf al-Assadi
15. Muhammad b. Abdullah b. Ahmad al-Sindi
16. Muhammad b. Ahmad
17. Fakhr al-Din b. Abu Bakr b. Muhammad b. Naimi
18. Ahmad b. Muhammad b. Ahmad b. Abdullah b. Yusuf al-Assadi
19. Siraj al-Din b. Abu Bakri Muhammad b. Ibrahim al-Salami
20. Ibnu al-Ma'ruf Ismail al-Bahri Shari'i b. Ibrahim [b. Abd al-Shamad al-Uqael al-Yamani al-Zubaedi] al-Jabarti, d. 1403]
21. Shihab al-Din Ahmad b. Abi Bakr al-Raddadi [d. 1418]
22. Muhammad b. Said al-Kubbani al-Tabri
23. Muhammad b. Mas'ud b. Sakika al-Ansari
24. Muhammad b. Ahmad b. Fadlillah
25. Fakhr al-Din Abi Bakri al-Aydarusi al-Adani [d. 1508]
26. Al-Said Abdullah al-Idrusi [b. 1513, d. 1582]

[This is the point where the central circles at the bottom of the Bira *silsila* begin.]

27. Al-Saiyyid al-Shaikh Amin Abdullah al-Aydarus [d. 1610]
28. Jamal al-Din al-Sayyid Shaikh Muhammad b. Abdullah [al-Aydarus Sahib Surat, b. 1561 in Tarim, moved in 1580 to Ahmadabad, Gujarat, d. 1620]
29. Al-Sayyid Shaikh Hasanji [b. Muhammad Hamid al-Raniri?]
30. Sayyid [Abu Hafs] Umar b. Abdullah b. Abd al-Rahman [Ba Shayban al-Tarimi al-Hadrami, d. 1656 in Bilgram]
31. Al-Shaikh Nur al-Din Muhammad b. Ali b. Hasanji b. Muhammad Hamid al-Raniri [b .c. 1590, hajji 1620, Aceh 1637–43, d. 1658]

[The last five names in the original *silsilah* are those of Bugis who settled in Bira.]

32. Haji Al-Shaikh-Julaij Ahmad b. Abdullah al-Bugisiya [Panre Lohe]
33. Al-Shaikh Abd al-Rahman b. Abdullah Lamatti [Panre Keke]
34. Shaikh Abd al-Jalil b. Abdullah Bulo-Bulo [Guru Toaya]
35. Shaikh Abd al-Basir b. Abd al-Jalil al-Bira wal-Bugisiya [Tu ri Masigi'na]
36. Shaikh Abd al-Fattah al-Hidayattullah Sharmallahu [To Daba]

Figure 3.3 The Qadiri *Silsila* of Bira

Q35. Shaikh Abd al-Basir ibn Abd al-Jalil al-Bira wal-Bugisiya (*To ri Masigi'na*, The One of the Mosque) [In a marginal note: He who built the mosque in Bira. His tomb is in the graveyard at Talaya in Bira Keke.]

Q36. Shaikh Abd al-Fattah al-Hidayattullah Sharmallahu (*To Daba*, Father of Daba)

Q32–35 are all identified as of Bugis origin. This is in accord with what the *kepala desa* of Bira, Amiruddin Said Patunuru, told Christian Pelras during an interview in 1984. According to Amiruddin, Shaikh Ahmad (Q32) married the daughter of the Arung of Lamatti and settled in Sinjai. One of his sons settled in Bone, while another, Shaikh Abd al-Rahman, settled in Bira after converting in to Islam (Pelras 1985: 111–112).

Although the realms of Lamatti and Bulo Bulo were ethnically Bugis, they had belonged to Gowa's sphere of influence since the Tangka River had been fixed as the boundary between Gowa and Bone in 1565. It would have been quite natural for Bugis from the former realms that had traveled overseas to study Islam to settle in the local boat-building center of Bira when they returned, where it would have been much easier for them to stay in contact with cosmopolitan networks of Islamic learning. They may have had another reason to keep their distance from Bone. If their teacher, Haji Ahmad, returned to Sinjai when al-Raniri fell from favor in 1643, he and his students would have found themselves in the middle of Gowa's campaign to crush the religious reforms initiated by La Ma'darammeng (see later). For whatever reason, Abd al-Rahman and Abd al-Jalil chose not to settle in their homelands, but to move south to Bira.

Bone had established itself as the most powerful of the Bugis states in the 1560s, the only one capable of resisting Gowa's domination of South Sulawesi. Fifty years later, Bone again proved itself the most resistant to Gowa's campaign to Islamize the whole peninsula. When Bone's newly installed ruler, La Tenrirua, finally accepted Gowa's demand that he convert in 1611, the *Hadat* removed him from office and replaced him with Arung Timurung La Tenripale. Gowa defeated an army raised by the latter, but allowed him to remain on the throne when he finally agreed to convert. He reigned until his death in 1630, when he was succeeded by his sister's son, La Ma'darammeng (Andaya 1981: 39).

In 1640, La Ma'darammeng began to enforce "a stricter version of Islam in his kingdom than had previously been known (Eerdmans n.d.: 18). He issued an edict forbidding anyone in his kingdom to keep or to use slaves who had not been born into slavery. All non-hereditary slaves were ordered freed or given wages for their labor"

(Andaya 1981: 39). Pelras and Mattulada add that he discarded the *bissu*, the royal transvestite priests who guarded the sacred regalia; prohibited gambling and the drinking of palm beer; and destroyed all the old shrines, places he viewed as sites of idol worship (*shirk*) (Pelras 1996: 142; Mattulada 1976: 55).

Little else is known in detail concerning the doctrines that inspired La Ma'darammeng. It is interesting to note, however, that he imposed his novel interpretation of Islam at the same time that al-Raniri was conducting his campaign against the teachings of Hamzah Fansuri and Shams al-Din in Aceh. Al-Raniri produced a huge volume of Malay writings setting out his own rigorous interpretations of Islamic doctrine according to which local rulers are subordinate in religious matters to cosmopolitan scholars and mystics like himself. These were diffused widely through island Southeast Asia. The Bira *silsila* analyzed earlier gives some support to the hypothesis that La Ma'darammeng's reforms were directly inspired by al-Raniri's teachings.

As we saw in chapter 2, the sultans of Tallo' and Gowa had taken over the doctrines of Hamzah Fansuri and Shams al-Din that glorified the sultan of Aceh as the Perfect Man and God's Shadow on Earth. al-Raniri's doctrines would have had an obvious appeal to a ruler of Bone whose nation had been at war with Gowa for almost a century, and which was now suffering under its arrogance in matters of religion as well. La Ma'darammeng issued his edict just one year after the end of the forty-four-year reign of the first Islamic sultan of Gowa, Ala al-Din (r. 1595–1639).

The doctrines espoused by La Ma'darammeng posed a challenge not just to the religious pretensions of the sultan of Gowa, but to the entire social order of Bone as well. The nobles of Bone rose in opposition to the decree that all nonhereditary slaves should be freed. They were led by La Ma'darammeng's own mother. Soon many Bone nobles were fleeing the kingdom and petitioning Gowa to intervene. Gowa took no action until La Ma'darammeng threatened to export his doctrines to the neighboring Bugis realms of Wajo', Soppeng, Massepe, Sawitto, and Bacukiki. This was a clear threat to Gowa's hegemony, and Gowa formed an alliance with Wajo' and Soppeng to attack Bone. La Ma'darammeng was defeated and forced to flee to Luwu' in 1643. Gowa appointed To Bala, a member of the Hadat of Bone, to serve either as *kali* or as regent, depending on the source. A Makassar noble, Karaeng Sumanna, was appointed as viceroy.

In 1644, La Ma'darammeng's brother raised a new army to restore Bone's independence. It was crushed by an alliance of Gowa, Wajo', Soppeng, and Luwu'. Bone was reduced from the status of a vassal of

Gowa to that of a slave. Many of its nobles were rounded up and brought to Gowa as hostages. La Ma'darammeng was captured and exiled to the village of Sanrangang in Marusu', where Gowa could keep a close watch on him (Abdurrazak Daeng Patunru [1960]: 53). There he founded a center for mystical studies in Marusu' and wrote a number of religious works that are still in circulation among Bugis scholars (Pelras 1985: 135 n.77; Pelras 1996: 188).

Bone's Ultimate Victory through Its Alliance with the VOC

Bone's salvation from the domination of Gowa came from an unexpected source, the Dutch United East India Company, or *Vereenigde Oost-Indische Compagnie* (VOC). The VOC was formed in 1602 when competing Dutch trading companies established a cartel to control the prices of spices imported from eastern Indonesia. The superiority of the naval power, global financial resources, and organizational structure of the VOC enabled it to subdue one Indonesian sultanate after another in the course of the seventeenth century.

After 1641, the only remaining Indonesian maritime powers of any significance were Banten and Gowa. Banten dominated native trade in the western part of the archipelago while Gowa dominated the east. With the occupation of Bone in 1644, Gowa had eliminated the last threat to its power in South Sulawesi. It then turned its attention to the spice trade in eastern Indonesia. A clash between Gowa and the VOC in eastern Indonesia was now inevitable. In 1660, the sultan of Gowa ordered the regent of Bone, To Bala', to conscript 10,000 laborers to build defensive works against a threatened Dutch attack. To guard against desertions, Bugis nobles were made to work alongside the conscripts and were held responsible for their work.

One of these nobles was La Tenritatta Arung Palakka (1633–1696). Arung Palakka was a minor Bugis noble who was born in the village of Lamatta in Soppeng in about 1633. Although he was a youth of no more than seventeen in 1660, he helped lead a rebellion among the Bugis working on the fortifications. They managed to escape to Bone and to raise an army. Gowa easily defeated it, however, and Arung Palakka fled to Butung with his family and a few followers. Not long after, they were forced to flee again to the protection of the VOC in Batavia. There they proved their usefulness to the VOC by participating in a campaign against the Minangkabau in Sumatra. Arung Palakka and his men finally got the chance to avenge themselves on Gowa in 1666 when they joined a Dutch expedition under the command of Cornelis Speelman.

In 1667, the allies were victorious and Gowa was forced to surrender and to sign the Treaty of Bungaya (Bulbeck 1990: 82). Under the terms of this treaty, the Dutch were allowed to build a solid fort at Ujung Pandang. This remained their headquarters in eastern Indonesia, with only brief interruptions, until 1949. The Bugis forces of Arung Palakka took over one of Gowa's three other principal forts at Bontoala'. This was to remain the west coast residence of the rulers of Bone until well into the nineteenth century. Many vassals of Gowa were transferred to the VOC by right of conquest. These included Marusu' to the north; Galesong and Polombangkeng to the south; Bantaeng, Bulukumba, and the ten little realms lying to the north of the cape of Bira, to the east; and the Sultanates of Sumbawa, Tambora, and Bima overseas.

Instead of tearing down their own remaining fortifications as required by the Treaty of Bungaya, the Gowanese began to reinforce them. This led to a second battle in 1669, when the fort at Somba Opu at the mouth of the Jeneberang River was sacked. Sultan Hasan al-Din was forced to abdicate in favor of his son, Sultan Amir Hamza (r. 1669–1674). When Amir Hamza died in 1674, he was replaced by his young brother, Sultan Muhammad Ali (r. 1674–1677). In 1677 a joint VOC and Bone force attacked and destroyed the last remaining Makassar fortification at Kale Gowa. Muhammad Ali was deposed and exiled to Batavia, where he died in 1681 (Andaya 1981: 168–169). When the VOC installed a third son of Sultan Hasan al-Din, Abd al-Jalil, as ruler of Gowa in 1677, he was a king in name only, unable to defend himself or his people against depredations of Bone or the VOC.

Summary

During the sixteenth and seventeenth centuries, the cosmopolitan networks of the Arabian Sea defined themselves in opposition to the religious pretensions of the Islamic rulers of gunpowder empires in monsoon Asia. They called on the legal teachings of the *ahl al-hadith* of the ninth century to resist the attempt by Shah Ismail of Iran, Akbar of India, Iskandar Muda of Aceh, Agung of Mataram, and Abdullah of Tallo' to revive the seventh- and eighth-century conception of the ruler as the political and religious heir of the Prophet. But they were also able to draw on the mystical practices that had developed in the Sufi *tariqa* during the twelfth century to claim a large measure of charismatic authority for themselves. It was to this authority that the Islamic rulers of northern India appealed when they were confronted by the depredations of the infidel Mongols in the thirteenth century. And it was to this authority that the Islamic rulers of Southeast Asia again appealed when they were confronted by the depredations of the infidel VOC during the eighteenth century.

La Ma'darammeng's attempt to implement the relatively egalitarian, cosmopolitan model of Islam prevalent around the Arabian Sea was ruthlessly suppressed by the emperor of Gowa during the 1640s. The hierarchical, locally centered model of Islam adopted by Sultan Abdullah lived on in the court of Gowa until Sultan Hasan al-Din abdicated in 1669. Arung Palakka acquired great prestige from his military accomplishments, but as an ally of the infidel VOC, his religious credentials were dubious at best. As we shall see, the heirs of both Sultan Hasan al-Din of Gowa and Arung Palakka of Bone were forced to turn to the cosmopolitan networks of *ulama* and *shaikhs* to restore a measure of charismatic authority to their regimes. And the most knowledgeable and powerful member of these networks turned out to be a humble Makassar of uncertain paternity called Yusuf.

Elite Mysticism and the Sultan of
Gowa, 1611–1705

After their defeat, the kings of Gowa abandoned their claim to religious supremacy. They invited cosmopolitan *shaikhs* who had studied in Mecca and Medina to serve as their chief religious advisers and to implement Islamic law as it was then practiced in the Ottoman Middle East. These *shaikhs* usually married local princesses and founded lineages from which religious officials called *kali* (from Arabic *qadi*) were recruited. The charismatic authority of these *kali* lineages complemented the traditional authority of the royal lineages, and the two lines tended to intermarry in each generation. This complementarity of *kali* and king may be seen as a sort of synthesis of the elitist and the populist models of Islam already discussed.

In South Sulawesi, the most prestigious of these charismatic lineages was founded by Shaikh Yusuf al-Maqasari (Drewes 1926; Feener 1998–1999). An analysis of a biographical narrative about Yusuf preserved by the *kali* of Gowa will reveal many layers of meaning. On one level, it provides an historical account of a crucial phase in the Dutch colonization of Indonesia. On a second level, it transforms Yusuf's life into a version of an ancient Austronesian myth in which the hero spends his life trying to reunite with his opposite sex twin. On a third level, it transforms Yusuf's travels around the Indian Ocean into a Sufi allegory of the mystic's quest to reunite with his Creator. On a fourth level, it embodies a political model of the relationship between the traditional authority of temporal rulers and the charismatic authority of spiritual masters in which the former are clearly subordinate to the latter. Finally, it articulates a religious model in which a mystical adept functions as a sort of universal shaman who is able to move back and forth between the world of the living and the world of the dead, serving as a mediator and intercessor between the two.

The Early Years of Shaikh Yusuf

Local Makassar tradition holds that Yusuf was born to a woman from the village of Moncong Lowe in 1626 (Ligtvoet 1880: 90; see also Abu Hamid 1994; Tudjimah 1987). Most traditions hold that his father was a *gallarrang*, or village chief of low rank. Some traditions give his father a more mysterious origin, identifying him with the immortal prophet al-Khidr. All agree that his mother was taken as a wife by the sultan of Gowa when she was already pregnant with Yusuf (Abu Hamid 1994: 79–89).

Gowa was already well supplied in the 1640s with Arabs who could teach many Islamic disciplines at quite a high standard. As a youth, Yusuf studied the Koran under a local Muslim called Daeng ri Tasammang and then moved on to Arabic language, *fiqh* (law), *tawhid* (theology), and *tasawwuf* (mysticism) under an Arab preacher who lived in the sultan's fort at Bontoala, Sayyid Ba Alawi ibn Abdullah al-Allamah al-Tahir (Azra 2004: 88). In 1641, Yusuf went to Cikoang to study under another Arab *shaikh*, Jalal al-Din al-Aidid. Al-Aidid was born in Iraq and began his travels by sailing to the Hadramaut in southern Arabia. There he was integrated into the network that the Hadramis had already established between the Arabian Sea and Indonesia. He arrived in Aceh early in the reign of the Sultan 'Ala al-Din Ri'ayat Shah (r. 1589–1604) and gathered a circle of disciples. When his teachings were securely established in Aceh, he moved on to Banten during the reign of Maulana Muhammad (d. 1596) and then to Banjarmasin in South Kalimantan. There he married the daughter of an exiled Makassar noble from Galesong in the kingdom of Gowa. He also made friends with Datu ri Bandang, the Sumatran missionary who had failed in his first attempt to convert the king of Gowa, Tunijallo' (d. 1590). When Abd al-Makmur finally succeeded in converting the kings of Tallo' and Gowa to Islam in 1605, he invited al-Aidid to visit his wife's homeland. Al-Aidid's interview with the Sultan Abdullah went badly, however, and he fled with his wife and children to Cikoang, a coastal settlement in Takalar. He was still teaching there in 1641 when Yusuf came to study with him, but he later moved on to the islands of Selayar, Butung, and, finally, Sumbawa, where he died.

Al-Aidid's eldest son, Sayyid Umar, remained in Cikoang and carried on his teaching. His descendents constitute a large part of the village's population to this day. As they have come down through many generations of his students, the teachings of al-Aidid placed an emphasis on asceticism, on an original conception of the end of the world, and on a particularly elaborate cult of the Prophet's birth. His school is known as the *Tarekat Bahar al-Nur*, "The Path of the Ocean of Divine Light" (Hamonic 1985).

After completing his studies with al-Aidid, Yusuf decided to travel to Mecca and Medina to seek knowledge from the greatest masters of his time (see map 3.1). In 1644, he received permission to go on *hajj* from the sultan of Gowa, Malik al-Said (b. 1607, r. 1639–1653). What we know of Yusuf's travels from this point derives primarily from a single manuscript "in which Yusuf lists the various *turuq* in which he was initiated and gives for each his *silsila* (spiritual pedigree)" (van Bruinessen 1991: 253 n.4).

Yusuf's first destination was the sultanate of Banten, then one of the leading centers of Islamic learning in Southeast Asia. Banten had been closely identified with Islam since the time of its foundation in the early sixteenth century. When Yusuf arrived in 1644, Abd al-Qadir was putting together another mission to obtain authoritative advice on religious matters. Yusuf may have been carrying a letter of introduction from his teacher al-Aidid, who had visited Banten in the early 1590s. He appears to have achieved a rapid acceptance in the royal court, and became friends with the heir to the throne, Pangeran Surya, who later reigned as Sultan Ageng Tirtayasa.

This mission was to consult the *Shaikh al-Islam* of Aceh, Nur al-Din al-Raniri, on certain questions about the teachings of Hamzah Fansuri (van Bruinessen 1995: 194 n.13; al-Attas 1986: 28). By the time this mission reached Aceh in 1645, however, al-Raniri had already fallen out of favor and returned to his home in Gujarat (see earlier). The mission must have caught up with him there, because one of only three works known to have been written by al-Raniri in Gujarat between 1644 and 1656 was composed in response to these questions of Sultan Abd al-Qadir.

Since we know that Yusuf al-Maqasari reached Aceh in 1645 after al-Raniri had left for India, and that Yusuf was initiated into the Qadiriyya by al-Raniri, it is quite likely that Yusuf joined this mission from Banten and accompanied it all the way to Gujarat. While in Gujarat, Yusuf also studied under al-Raniri's own teacher, Sayyid Abu Hafs Ba Shayban (d. 1656). From Gujarat, Yusuf sailed to Nuhita in Yemen, where he was initiated by Muhammad Abd al-Baqi al-Mizjaji

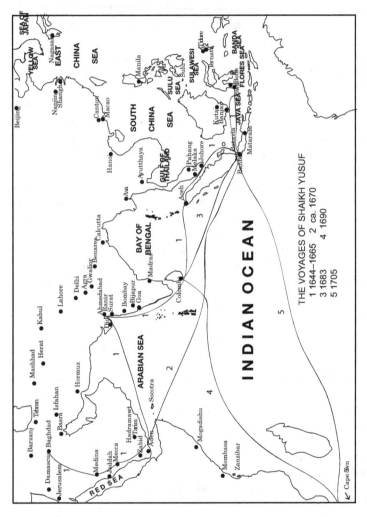

Map 3.1 The Travels of Shaikh Yusuf, 1644–1705

into a branch of the Naqshabandiyya that strongly supported Ibn al-Arabi's doctrine of the Unity of Being (van Bruinessen 1991: 255). In Yemen, Yusuf was also initiated by Sayyid Ali into the Alawiyya, an order that was usually limited to descendents of the founder and that was largely confined to the Hadramaut (see Knysh 1999 for an account of Sufism in medieval Yemen).

Yusuf continued over land to Medina, where he was initiated into the Shattariyya by Shaikh Ibrahim al-Kurani (d. 1689), who was also a teacher of Abd al-Rauf of Aceh (see figure 3.1). Shaikh Ibrahim engaged in fierce polemics in defense of Ibn al-Arabi's doctrines against the followers of the Indian Ahmad Sirhindi (1564–1624, see Rizvi 1983 II: 319–347). Two manuscripts of al-Jami's *al-Durrah al-Fakhira* copied by Yusuf while studying under Ibrahim al-Kurani in Medina survive. They were written in 1655 and 1664, so we may infer that that he was engaged in at least nine years of intensive study of mysticism in Medina (Heer 1979: 13, 15).

Given the emphasis of all his teachers on the doctrines of Ibn al-Arabi, it is not surprising that Yusuf went on to visit his shrine in Damascus. There he was initiated into the Khalwatiyya by the Imam and Khatib of the mosque associated with the shrine, Abu Barakat Ayyub al-Khalwati al-Quraishi. In South Sulawesi, Yusuf is most strongly associated with the Khalwatiyya, even though his writings draw mostly on Naqshabandiyya sources (van Bruinessen 1991: 253–256).

Ottoman Orthodoxy in Banten and Gowa, 1670–1705

Yusuf's friend, Sultan Ageng, came to power in Banten in 1651. He immediately sought the blessing of the Grand Sharif of Mecca, as his grandfather had done in 1638. When Yusuf returned to Indonesia in the late 1660s, Sultan Ageng persuaded him to settle in Banten instead of returning to Gowa. Sultan Ageng gave Yusuf his daughter in marriage and appointed him the *Shaikh al-Islam* of Banten. Yusuf proceeded to implement the religious practices current in the Ottoman provinces of the Middle East (Kathirithamby-Wells 1970: 52).

When Sultan Abd al-Jalil of Gowa began his reign in 1677, the Makassar people were in desperate straits both at home and abroad. One of Abd al-Jalil's first actions was to write to Shaikh Yusuf in Banten, begging him to return to his homeland and serve as tutor to the crown prince. Yusuf refused, but sent in his stead a blind Makassar he had met in the Hejaz and had brought back with him to Java. His full name was Abu al-Fath Abd al-Basir al-Darir, but he was better known following his death as *Tuan Rappang*, the Lord of Rappang, after the place where he lived out his final years in northern South Sulawesi (Matthes [1885] 1943: 391).

We are fortunate to have an account of conditions in Gowa during the era of Tuan Rappang as seen through the eyes of two young Makassar princes (Pelras 1997, 1998). The information these two princes provided a French Jesuit, Nicolas Gervaise, in 1686 casts an interesting light on the state of Islam at the time of Tuan Rappang, at least as it was viewed by the anti-Dutch faction of the Makassar nobility. As a priest, Gervaise took a particular interest in the Muslim clergy in Gowa. He distinguished three "orders" of "Agguy" (*Hajji*). The first order he called *labes*, the Makassar term for the *muezzin*, the mosque official in charge of the call to prayer. According to Michael Feener, *labes* is derived from Tamil (personal communication). The second order he called *santari*, or religious students. These were celibate officials who lived inside the mosque and were in charge of maintaining the mosque and its library. They shaved their heads, wore a plain sarong of white linen, and subsisted on daily donations of alms. The third order he called *tuans*, lords. He wrote that this rank was conferred by the Grand Mufti in Mecca, and that the *tuan* who was closest to the king was "the Patriarch and Primate of the Kingdom; nor does he acknowledge any one above him, but the Grand Mufti of Mecca" (Gervaise [1688] 1701: 155). The preeminent *tuan* at the time must have been none other than Tuan Rappang. According to Gervaise, the *tuans* dressed in the Turkish style with long white robes, and wore turbans when leading prayers. As to the general piety of the population, Gervaise had this to say:

> And now it is not to be imagin'd, with what exactness the Macasarians acquit themselves of the Duties enjoyn'd by their new Religion: they would not miss of the meanest Holydays which it prescribes, without signalizing their Devotion, every one in particular, by some Good Work or other, of Supererogation; the neglect of a Bow, or any slight Washing, is look'd upon by them as a considerable Crime. Some of them, out of a mere sentiment of Repentence, abstained all their Lives from drinking Palm-Wine, tho' it be not forbidden by the Law. And some there are, that will rather dye for Thirst, than Drink so much as a Glass of Water, from Sun-rising to Sun-setting, during the whole time of their Lent. More that this, they are far more devout that all the other Mahometans; for they observe an infinite number of Ceremonies that are not in use among the Turks, nor among the Indian Mahometans; because they believe them to be practis'd at Mecca, which they look upon as the Center of their Religion, and the Pattern which they ought to follow. (Gervaise [1688] 1701: 133)

It is clear that by the 1680s the attempt to treat Islam as the product of local revelations and sources of power had lost ground to a fervent cosmopolitanism, in which Mecca was the source of legitimacy for all the leading religious authorities and for correct religious practice. The sultans of Banten and Gowa had been so weakened by the VOC that they did not pretend to have the same degree of religious authority their own predecessors had at the beginning of the seventeenth century. They deferred instead to cosmopolitan *shaikhs* such as Tuan Rappang and Shaikh Yusuf, who helped impose the model of Islamic orthodoxy they had learned during their time in the Ottoman-dominated Hejaz.

Shaikh Yusuf and Sultan Ageng of Banten developed a fervently anti-Dutch foreign policy. They supported a rebellion of Minangkabau living near Melaka in 1677, an insurrection in Ambon in 1680, and an uprising in west Sumatra in 1681. The Dutch began to look for internal cleavages in the sultanate that they could exploit for their own ends. The heir-apparent to the throne of Banten, Abd al-Qahhar had spent another two years in Mecca from 1674 to 1676. During his absence, Sultan Ageng appointed another son, Pangeran Purbaya, heir apparent. When Abd al-Qahhar returned from Mecca, a faction favorable to the VOC formed around him and power in Banten was increasingly divided between the "Old Sultan," Ageng, and the "Young Sultan," Abd al-Qahhar, now known as Sultan Haji. In 1682, the VOC marched on Banten to support Sultan Haji. Sultan Ageng fled into the interior with Shaikh Yusuf and his favored son, Pangeran Purbaya. They eluded the VOC forces for almost a year, but were finally tracked down with the help of a gang of escaped slaves under the leadership of Surapati. Sultan Ageng was imprisoned by his son, Sultan Haji. Shaikh Yusuf was taken into custody by the VOC. He had initiated many Makassar exiles into the Khalwatiyya order over the past twelve years, and these followers were allowed to return to Gowa (van Bruinessen 1995: 182). In 1683 Yusuf's sons established their own Khalwatiyya center in Marusu' which came to be associated primarily with the high nobility Gowa and Tallo'. It was located just six kilometers away from the Sufi lodge founded by La Ma'darammeng after he was deposed as king of Bone in 1640 (Ligtvoet 1880: years 1714 and 1715).

Shaikh Yusuf was held captive in Batavia at first, and was then sent into exile in Sri Lanka accompanied by some of his family and close followers. Yusuf spent ten years in Sri Lanka. Azra sees this exile as a blessing in disguise, since it allowed Yusuf to avoid the distractions of political struggle and to devote himself exclusively to religion

(Azra 2004: 98). Yusuf spent his time composing numerous works on mystical topics and corresponding with his followers in Indonesia (Voorhoeve 1980: entries numbered 41, 52, 82, 129, 148, 203, 246, 279, 341, 347, 354, 408, 461, 463, 467).

Yusuf was in close contact with several Indian Sufis in Sri Lanka, including Sidi Mailaya, Abu al-Ma'anni Ibrahim Minhan, and Abd al-Siddiq bin Muhammad Sadiq (Azra 1992: 440, 450). He abandoned Ibn al-Arabi's doctrine of the Unity of Being, which was favored by most of his earlier teachers, and went over to Sirhindi's doctrine of the Unity of Witness, which was currently in vogue in the court of Aurangzeb (r. 1658–1707; Abu Hamid 1994; Azra 2004: 104). Yusuf never abandoned Ibn al-Arabi's other doctrines that the creation of the world depends on God's desire for the Perfect Man and that the student must follow the guidance of his chosen master implicitly even when the latter appears to transgress the *shariah* (Azra 2004:107–108). Hamka claimed that Emperor Aurangzeb took personal notice of Yusuf's presence on the island, and warned the Dutch to treat him well (see Hamka 1963: 46–47; but note that Azra could find no record of this in the Dutch or Indian archives, Azra 2004: 104).

After Sultan Haji came to power in Banten in 1683, continued resistance to the VOC in Southeast Asia depended on charismatic leaders acting in the name of cosmopolitan Islam. In 1685, a Minangkabau called Ahmad Shah ibn Iskandar declared a *jihad* against the VOC throughout Indonesia. He assembled a fleet of 300 vessels manned by 4,000 Makassar, Minangkabau, and Malays. In 1686, Surapati, the former slave and erstwhile lieutenant in the Dutch army who had helped capture Yusuf, came under Ahmad Shah's influence and led a separate uprising in east Java. In 1687 the British also began to support Ahmad Shah's movement from their base in Bengkulen. For the next nine years Ahmad Shah fought a guerrilla campaign in Sumatra against the Dutch and their Bantenese allies (Kathirithamby-Wells 1970). In 1687 a Gujarati emissary of the Mughal Emperor Aurangzeb was arrested for inciting rebellion in Ambon.

The Dutch thus had good reason in the 1680s to fear the growth of an Islamic resistance movement that spanned the Indian Ocean. This movement also made local rulers who had signed treaties with the Dutch fearful about their own fate should the rebels succeed. The tenure of Sultan Abd al-Jalil of Gowa was particularly precarious because of the way the Dutch had openly installed him in 1683 after banishing his brother Muhammad Ali. Abd al-Jalil did not manage to secure the recognition of Gowa's *Bate Salapang*, Council of

Nine Electors, until 1689. It was then that he sent an official request to have Shaikh Yusuf released from exile in Sri Lanka and returned to Gowa. Hartsink agreed to this request but was overruled by a special commissioner sent from Batavia, Dirk de Haas, who noted that Arung Palakka himself was steadfastly opposed to the return of such a prestigious Gowan who might become the focus of opposition to his own hegemony (Andaya 1981: 276).

In 1691, de Haas reported that the royal family of Gowa was beginning to claim close kinship ties to Yusuf, and even that he was a half brother of Sultan Abd al-Jalil (Andaya 1981: 277). Far from acceding to the latter's continual requests that Yusuf be returned to Gowa, however, the Dutch decided to move him even further away. In 1693 he was sent to the Cape of Good Hope in South Africa, where he died in 1699.

In the absence of a living *shaikh* to legitimate his authority, Sultan Abd al-Jalil turned to a dead *shaikh*. In 1701, he began to sponsor royal processions to the tomb of Datu ri Bandang, the Sumatran *shaikh* responsible for converting Gowa and Tallo' to Islam a century earlier. Sultan Abd al-Jalil also established a connection to the spirit of Shaikh Yusuf through a series of dreams that were carefully recorded in the royal diaries. In 1703 he requested the return of Yusuf's mortal remains, and this request was granted. According to the *Diary of the Kings of Gowa*, Yusuf was re-entombed in Lakiung near the graves of the Karaengs of Gowa on April 5, 1705 (Ligtvoet 1880: 176). His Gowanese wife, Daeng Nisanga, was later reburied next to him. Abd al-Jalil granted Yusuf's family freedom from enslavement, tolls, levies, taxes, and feudal service. He inaugurated annual processions to the joint tomb of Shaikh Yusuf and Daeng Nisanga (Cense 1950: 53; see Goldzihir 1971 for a classic account of Islamic "sainthood").

Sultan Abd al-Jalil also linked the royal house of Gowa to Yusuf by arranging marriages with his descendents. According to the *Diary of the Kings of Gowa*, Abd al-Jalil's maternal grandson and heir, Karaeng Anamoncong, married Sitti Labibah, one of Shaikh Yusuf's daughters by Kare Kontu, in 1706 (Ligtvoet 1880: 178). This Karaeng Anamoncong later served as Sultan Ismail of Gowa (1709–1712) and of Bone (1720–1724). In 1721, Sultan Najm al-Din (r. 1723–1729) of Tallo' was betrothed to Zainab, the daughter of Labibah and Sultan Ismail, and was married to her in 1724. The intermarriage between these royal houses and the lineage of a *shaikh* set a precedent that was followed all over South Sulawesi. In villages such as Ara and Bira, this practice took the form of intermarriage between the lineage of *kali* that descended from Haji Ahmad and the lineage of *karaeng* that descended from the royal ancestor, Karaeng Mamampang (see chapter 6, figure 6.1).

The Riwayat Shaikh Yusuf

Having set out the facts of Shaikh Yusuf's life as we know them from historical sources, I want to turn now to an analysis of a narrative account of his life that was preserved by the royal house of Gowa. This narrative was first published in 1933 by Nuruddin Daeng Magassing. It was taken from a manuscript in the possession of the *kali* of Gowa, Haji Ibrahim Daeng Pabe (Nurdin Daeng Magassing [1933] 1981: 105; Hamka 1961, 1963: 37; Feener 1998–1999). An analysis of this text helps explain why the sultans of Gowa were so eager to claim Yusuf's charismatic authority, both during his life and after his death. The transliterated text is about sixty-pages long and may be summarized as follows:

1. The Dampang of Ko'mara' sees a bright light in his field at midnight, with an old man in it. The old man agrees to guard the field for him, then asks to become the servant of the Gallarrang of Moncong Loe. He marries his daughter.
2. The Karaeng of Gowa sends a messenger to find out why the Gallarrang of Moncong Loe has not come to an audience. The messenger sees the old man and his wife and reports on her beauty. The Queen demands to see her. The old man gives his wife to the king and disappears.
3. The daughter of the Gallarrang is pregnant with Yusuf when she goes to the court to marry the Karaeng. Light is seen to emanate from her womb, and the chanting of *dhikr* is heard from it. Yusuf is born in Parang Lowe in Tallo'. The King adopts him, and the Queen gives birth to Siti Daeng Nisanga, making them twins of a sort. Yusuf learns the Koran, grammar and logic from Daeng Ritasame. He is circumcised at age twelve along with Daeng Nisanga. [Female "circumcision" among the Makassar involves only a small incision in the clitoral sheath, not the removal of any tissue.] She declares her love for him. He argues he is too low in rank to reciprocate, since his mother was only a Gallarrang.
4. I Dato' ri Pangengtungang and I Lo'mo' ri Antang go with Yusuf to study with local *shaikhs* at Bulu Saraung, Lanti Mojong and Bawakaraeng. They learn their first esoteric knowledge, *ilmu*, there, which gives them the power to light fire from water. The *walis* tell Yusuf he must go to the Imam of the Shafiites in Mecca if he wants to learn more *ilmu*.

5. Having learned this much *ilmu*, Yusuf decides he is ready to ask for Daeng Nisanga's hand in marriage. The King contemptuously rejects him because he is too low in rank. Yusuf swears not to return to Gowa until his Sufi path is complete—that is, not until after his death. After Yusuf leaves, the King's councilors read to him from a *lontara* manuscript that high rank can be achieved through acquisition of *ilmu*, the display of bravery and the accumulation of wealth. The king changes his mind and calls Yusuf back, but Yusuf refuses to return. He has been too deeply insulted. The princess is sent to him in Kampong Beru to plead with him to stay, but after forty days he sails for Batavia without her.

6. In Batavia, Yusuf takes ship for Sri Lanka. The Captain dislikes him so Yusuf tilts the boat with a *dhikr* to demonstrate his power. Three days from an island, Yusuf separates his soul, *nyawa*, from his body and is buried at sea. Three days later, they find him alive when they land on the island. His *ilmu* is already so great that he is able to reverse the passage from life to death. Then the Prophet Khidr appears in disguise and asks the crew to hold his body while he dies. Only Yusuf does so, and stays with the rotting corpse for many days while the others sail away. Khidr returns to life and gives him further *ilmu* by spitting in his mouth. Yusuf catches up to the boat by walking across sea. Lo'mo' ri Antang dies and is buried at sea. When the boat is becalmed, Yusuf has a *nun* fish tow it to Jeddah. Lo'mo' ri Antang appears alive on the shore. Yusuf has now overcome death three times. But Yusuf tells Lo'mo' ri Antang his time on earth is over and sends him back to the other world.

7. Yusuf walks to Mecca protected only by his three *keris*, daggers. When they refuse to open door of the mosque for him because prayers have started, Yusuf causes the whole Kaba to tilt over on its side. They then accept him as the Shaikh foretold in the Koran by Ali as coming in the year 1110 A.H.

8. The Imam of the Shafii sends him to the Imam of the Maliki, who sends him to the Imam of the Hanbali who sends him to the Imam of the Hanafi. In this way he acquires the entire corpus of the *shariah*, the knowledge of the external laws of Islam held by the *ulama*.

9. The Imam of the Hanafi then sends him to see 40 shaikhs who have been dead for 225 years. The 40 *shaikhs* send him to the Teacher of the *shaikhs*, Abi Yazid al-Bustani, who has been dead for 500 years. He blows blessings, *berkat*, in his mouth and sends him to the King of the shaikhs, Abd al-Qadir Jilani, who has been dead for 750 years. He tests him by making him gouge out his own eyes, and then tells him he is equal in rank to him. In this way he acquires the entire corpus of the *tariqa'*, the knowledge of the internal mystical meaning of Islam held by the *wali*.

10. Yusuf then visits the tomb of the Prophet Muhammad. He must trick the gatekeepers into letting him through the seven portals that guard it, symbolic of the Seven Grades of Being. In the process he twice forgets himself utterly, achieving the state of *fana*, the loss of self as one is absorbed into the absolute. The Prophet Muhammad names him *Qutb al-Rabbani wal Arifin al-Samdani*, Pole of the Rabbani and Master of the Samdani. The Prophet sends him to Qasm al-Sirri, who has been dead for 800 years. Qasm sends him up a red river for forty days and a waterfall for another forty. The Prophet Yusuf protects him from Iblis on the way. From the top of the mountain, Yusuf throws himself off, but is carried up by the water to heaven after being unconscious for an hour. There the Tree of Fate asks him to eat from it so it may be blessed. Yusuf meets the Prophet Muhammad in heaven who sends him back to earth. When Yusuf protests, Muhammad opens a pinhole to Hell and the smell alone is enough to make Yusuf obey. On the way down, he disobeys Muhammad's orders and opens his eyes, causing him to fall onto the field of Mahsyar, whence he is rescued. He thus acquires all the knowledge of the Prophets concerning the Afterlife and Judgement Day when the world of creation will come to an end.

11. Back in Mecca, Yusuf becomes a famous teacher of the *wahdat al-wujud*, impressing many Arab *shaikhs*, including Shaikh Masym and Sayyid Muhyi al-Din Taj al-Kabdi al-Hadramawti. He makes the transition from pupil to master. The Sultan of Dima [Bima] hears of his growing fame and sends word offering his daughter in marriage. Yusuf refuses, quoting the words of the Karaeng of Gowa about his low rank. He sends Sayyid Muhyi al-Din in his place, who is known thereafter as Tuan ri Dima.

12. Yusuf sets out to see Rum (Istanbul), the political center of the Islamic world. The Prophet Musa stops him and describes it for him instead, saying his destiny is not there but to leave descendants in Banten. Yusuf begins to visit Banten regularly by walking there from Mecca in a few hours. When he saves the realm from a tidal wave, the Sultan offers his daughter in marriage. Yusuf finally overcomes his scruples about rank and agrees. His wife gives birth to Muhammad Abd al-Kadir, later known as I *Tuan ri Takalara'* "The Lord of Takalar," and a daughter called Siti Hanipa.

13. Abd al-Basir *Tuanta ri Rappang*, Lord of Rappang, studies under Yusuf in Mecca and goes with him to Banten. The Karaeng of Rum hears of Yusuf's marriage and sends a letter to the Karaeng of Gowa. The Karaeng of Gowa sends Daeng Mallolongang, Karaeng Rappocini' and Daeng Kare Nyampa to Banten to ask Yusuf to return. Yusuf again refuses and sends Tuan Rappang back in his place, along with Daeng Kare

Nyampa under the new name of Tuan I Daeng ri Tasammeng. They enforce rigid Islamic laws there.

14. Yusuf saves Banten from a famine that is rotting all foodstuffs and converts the local Dutch Governor to Islam. He thus shows the supremacy of Islamic knowledge over even the powerful Dutch. The Governor General sends this renegade Governor to Ambon and exiles Yusuf to Batavia. When Batavia is struck by an earthquake, Yusuf is sent to Ceylon. The Karaeng of Ceylon gives him his daughter in marriage, I Pipa. Their son is called Shaikh Alam, Master of the World.

15. Yusuf goes to the Cape of Good Hope and gains followers there. He now appears regularly in four places: Mecca, Banten, Ceylon, and the Cape. His wife in Banten dies and he marries her younger sister. She gives birth to Siti Habiba and Muhammad Abdullah Ance Daeng I Tuan Beba'.

16. Yusuf's principal male disciples each possesses a distinct *ilmu*: Muhammad Abd al-Kadir I Tuan ri Takalara' has The Unfathomable Sea; Muhammad Abdullah Ance Daeng I Tuan Beba' has the Limitless Sky; Sayyid Muhyi al-Din Taj al-Kabdi al-Hadramawti has the Rudderless Boat.

17. Each of Yusuf's daughters marries a descendent of the Prophet, so that none of his descendants fall below the rank of Sayyid on their father's side.

18. Yusuf dies in Banten after a seven-day fever. He tells his followers not to bury him until the Gowanese come. The Gowanese demand the body, and Yusuf tells the Bantenese to keep his shirt in a coffin. After three days they find his corpse inside and bury it. Yusuf sings a *kelong*, a four-line poem, to make the day longer. The Gowanese sail east to Gowa but end up west in Sri Lanka. The Sri Lankans open the coffin and find it full of large worms. They leave, but Yusuf has them fetched back and now the coffin contains his body full of light and beauty. Again there is a quarrel over the body and Yusuf tells the Sri Lankans to keep his hat for three days. It becomes his body and they bury it after Yusuf sings another *kelong*.

19. Seven days later, the Gowanese find themselves at the Cape. They open the coffin and find it full of white sand. They leave and are fetched back, and now it has his body. There is another fight, and Yusuf tells them to keep his prayer beads, which turn into a corpse that is buried after another *kelong*.

20. The coffin arrives in Gowa and is found to be full of water. The nobles all drink some. The next morning it is full of white sand, a *tasbih*, a *selawat* and a *kitab tarikat*. Yusuf's original fiancée is sent for and becomes pregnant. Separated in childhood, the two pseudo-twins can only be united in death. A final *kelong* is heard, and he is buried, a rainbow covering his grave. His wife

gives birth to Muhammad Maulana, father of Karaeng Tumenanga ri Tampa'na. (Synopsis of Nurdin Daeng Magassing [1933] 1981)

Analysis and Conclusion

At a surface level, *Riwayat Shaikh Yusuf* records a Makassar version of early colonial history in which the Dutch number as one among many earthly powers that are forced to acknowledge the superiority and universality of the Islamic religion. At a deeper level of meaning, it is a typical Austronesian myth in which a pair of twins are separated in childhood and reunited only after a protracted series of adventures. This is given a typically Makassar twist when the separation is ascribed to the difference in social rank between the male and the female. In the great pre-Islamic epic of the Bugis people, the *I La Galigo*, Sawerigading is drawn to his twin sister, We Tenriabeng, but is forbidden to marry her. After a lifelong search, his conflict is resolved when he marries We Cudai. She is the daughter of his mother's twin sister, who has the same name as his own twin sister We Tenriabeng. The La Galigo myth resolves the issue of incest by substituting the most closely related woman possible (Pelras 1996: 88–89; Gibson 2005: 63–77). This story is reminiscent in turn of the "Tale of Panji," found all over Indonesia (Rassers 1922; Ras 1973; Gibson 2005: Chapter 4).

The *Riwayat* raises and resolves a similar issue by making Daeng Nisanga not an actual twin, but a stepsister of Yusuf. Yusuf's life is structured by his quest to reunite with this stepsister, Daeng Nisanga. Although they have different biological mothers and fathers, they are born at the same time and place to the same legal father. Later, they enter Islam through circumcision at the same time.

As pseudo-twins Yusuf and Daeng Nisanga are irresistibly drawn to one another. Their marriage is prevented not by the law against incest, but by the sultan's refusal to accept Yusuf as equal in rank due to his low birth. The contrast between Yusuf's origins and that of the sultan of Gowa could not be greater. Yusuf is born of a low-ranking mother by a father of unknown origin. The sultan has a royal pedigree stretching back to the founding of the kingdom by a heavenly being in the thirteenth century. Yusuf's quest originates in the sultan's ignorance of the fact that high rank can be achieved through knowledge, bravery, and wealth, and Yusuf has already demonstrated the possession of superior knowledge. Out of shame, Yusuf swears never to set foot in Gowa again until after his death, foretelling the return of his body from Cape Town.

At a third level of meaning, the *Riwayat* is a mystical allegory in which the earthly separation from a Beloved woman stands in for the mystical separation from the Divine. In much mystical poetry, separation from the Beloved is a metaphor for the sundering of the Creature from his Creator. Yusuf's departure from Gowa in his youth is an allegory for the descent of the Creature into the material world. His ultimate return to Gowa and reunification with his Beloved can occur only after a lifetime of seeking Islamic knowledge.

One attraction of the cosmopolitan Islam taught by the wandering Sufis of southern Arabia and India derived from its promise to allow an individual to escape the confines of a theoretically fixed social hierarchy, to move up and down the ladder of power, wealth, and prestige at will. Obscure origins could even be an advantage in such a world, for one could always claim descent from a lineage of *shaikhs*. It should be noted that the ruler is never referred to by a universal religious title such as caliph, but by the local political title *karaeng*. There is no hint in this text of the idea that the *karaeng* is himself a mystical adept, much less a caliph or Perfect Man.

The *Riwayat Shaikh Yusuf* thus adds a charismatic Islamic twist to the traditional Austronesian myth of the wandering hero. He sets out from an imperial center in search of a source of knowledge and power superior to that derived from royal descent. He travels to the center of the entire cosmos, Mecca. There he ascends an infinite hierarchy of spiritual knowledge and is acknowledged as among the chief of the *shaikhs*. He founds sites and lineages of supernatural blessing all around the Indian Ocean by marrying princesses, fathering children, and being buried in four separate places: Banten, Sri Lanka, Cape Town and, finally, Gowa. It is only after his fourth and final burial in Gowa that Yusuf impregnates his "step-twin," Daeng Nisanga, from beyond the grave. The original unity and equality experienced by opposite-sex siblings in the womb is recoverable only after a radical separation in life, and is perfected only in the tomb.

At a fourth level of meaning, the *Riwayat* is a political commentary on the superiority of the charismatic power *shaikhs* to the temporal power of rulers. Despite his low social rank as defined in local Makassar terms, Yusuf is clearly endowed with charismatic powers even before his birth. The mysterious stranger who fathered him represents the power of prophets, a universal source of rank and power that transcends all local social systems. One may see in him a trace of all the cosmopolitan *shaikhs* who traveled around the Indian Ocean fathering children by local wives. In this respect, Yusuf's travels represent the gradual fulfillment of a universal destiny as his innate charismatic powers are expressed on an ever-broader geographical stage.

Shaikh Yusuf's life provided three distinct ways the model of the cosmopolitan shaikh could be applied to actual political situations. First, Sultan Abd al-Jalil's appointment of Tuan Rappang as Shaikh al-Islam helped establish the cosmopolitan Islam of the Hejaz as orthodox in Gowa, severing them from local rituals. Abd al-Jalil later created a new state cult centered on the tombs of Datu ri Bandang and Shaikh Yusuf. Second, Shaikh Yusuf's encouragement of military resistance to the VOC between 1670 and 1683 provided a charter for religiously motivated resistance movements in later centuries. The kind of esoteric religious knowledge he acquired by wandering across the Indian Ocean functioned as a sort of wild card in the increasingly oppressive mercantilist order imposed by the VOC. Men such as Yusuf of unknown origin could acquire charismatic power by traveling outside the boundaries of familiar social space and return to claim a new place at the top of society. The fluidity of their social identity enabled them to use the global prestige of Islam to unite local ethnic groups against European power. Third, Shaikh Yusuf integrated generations of Makassar into a cosmopolitan world in which no regional political order could claim a monopoly on religious truth. His mystical writings and the branches of the Khalwatiyya he founded in Java, Sri Lanka, and South Africa created enduring linkages between South Sulawesi and the whole Indian Ocean.

On a final level of meaning, the esoteric knowledge Yusuf gains during his travels turns him into a kind of universal shaman. It gives him the power to conquer space and time, life and death. All kings are ultimately forced to acknowledge his superiority. Indeed, by the end of his life, Yusuf's mastery of mysticism has enabled him to transcend not just the cultural distinctions made by members of his own ethnic group, but by human beings of all times and places. His moral triumph over the Europeans is expressed through his conversion of the Dutch governor of Banten to the universally true faith. He has transcended the quest for temporal power completely. And that is not all, for just as he overcomes all cultural boundaries, he overcomes all natural distinctions as well, moving instantaneously back and forth between life and death and from one side of the globe to the other.

The source of all this power lies in mastering the Unity of Being, a mystical state in which all lower-level conceptual and perceptual distinctions are confounded. The source of all Truth, Power, and Reality lies outside the created world altogether. The path to this source leads outside the familiar local social and political structure to a sacred center that lies at the edge of all the great empires of the day, the Kaba in Mecca. A great *shaikh* who can follow this path to its ultimate terminus gains a power beyond the reach of the mightiest temporal ruler. The power thus achieved outlives the temporal life span of the *shaikh* and gains an eternal presence. Thus Yusuf is able to freely converse with all the great *shaikhs* of the past, who coexist at a deeper level of reality. Like the Buddha, the great *shaikhs* achieve ultimate enlightenment during their lifetimes but then return to temporal reality to teach others and to provide a source of blessings for those still ensnared by the material world. Their power is directly proportional to the time they have spent on the other side of the boundary between life and death. This is clearly indicated by Yusuf's successive introductions to *shaikhs* who have been in the grave for ever longer periods.

In the case of Islamic *shaikhs*, these blessings can be tapped through *siara*, visiting the site of their mortal remains, and through their spiritual and biological descendants. The tombs and heirlooms left behind by transcendent figures such as Yusuf serve as points of access to a divine realm in which ordinary constraints of time, space, and social hierarchy are suspended. By performing rituals of homage to the *shaikhs* at their tombs, one can enter into communication with them leading to the acquisition of mystical knowledge, or into a patron–client relationship in which the granting of a favor is repaid by the fulfillment of a vow, *nazar*. Yusuf takes care that his tombs, disciples, and children are left behind in every corner of the Islamic world, from the Cape of Good Hope to his land of origin, Gowa.

It is important to note that it is not just the tomb of Shaikh Yusuf that is an object of veneration in South Sulawesi today. It is the *joint* tomb of Yusuf and his "sister"/wife, Daeng Nisanga, that is the object of pilgrimage today, especially by newly weds desirous of obtaining fertility. In a manner that recalls the Merina of Madagascar as analyzed by Bloch, the bisexual tomb of the ancestors replaces the house of the living as the source of blessing and fertility (Bloch 1971, 1986).

The tombs of Shaikh Yusuf and Daeng Nisanga are objects of veneration not just for a particular ethnic group. Together with Yusuf's other three tombs in South Africa, Sri Lanka, and Java, these tombs provide blessings to the entire Islamic community spanning the Indian Ocean. The merging of Austronesian with Islamic symbols of the unity and equality of all humans in the afterlife has transformed the ancestors of a localized bilateral descent group into the *shaikhs* of the *umma*.

Chapter 4

Islamic Martyrdom and the Great Lord of the VOC, 1705–1988

In the lands that were transferred to the VOC by the Treaty of Bungaya in 1667, the legitimacy of the local rulers collapsed almost completely. They were forced to pay tribute and homage to the VOC governor in Ujung Pandang, who had final say over who would succeed them in office. This was a role once played by the sultan of Gowa. But unlike the sultan, the governor could never acquire the traditional authority that came from intermarriage with local royal houses, nor the charismatic authority that came from sponsoring Islamic rituals and enforcing Islamic law. In the outlying areas, real power fell into the hands of corrupt junior merchants who made most of their profits from the trade in mostly female slaves.

The illegitimate nature of VOC power tainted the authority of all local rulers who were forced to acknowledge it. This was all the more true in the territories that had become vassals of the VOC in 1667, where local rulers served at the pleasure of a VOC "senior merchant" (*opperkoopman*) often a young man at the beginning of his career. With no legitimate political authority to appeal to, many Muslims abandoned all hope of bringing the temporal order back into line with Islamic teachings. They dedicated themselves instead to a mystical path through which they cultivated an indifference to life in this world. When the time came, they were happy to find salvation by dying in battle with the forces of evil. To grasp just how evil VOC power could appear, it is necessary to understand how closely intertwined it became with the slave trade in South Sulawesi.

Makassar nobles were famous throughout Indonesia for exacting violent retribution whenever local Muslim rulers insulted the honor of their women. When a European infidel insulted a Makassar woman, the man responsible for her was expected to redeem her honor by defeating the perpetrator in battle. Since Europeans enjoyed the protection of disciplined VOC troops, individual attempts to achieve satisfaction often resulted in death. Europeans considered such acts a form of madness, or "running *amok*." Makassar often interpreted them as a form of *jihad*, or righteous struggle in defense of religion. Those who died in single combat against the Dutch army were seen as *shahid*, martyrs who went straight to paradise after their death on the battlefield.

This is the theme of a favorite work of Makassar oral literature, the *Sinrili' Datu Museng*, which is often recited over several nights before a wedding. In this epic, Datu Museng begins life as a poor shepherd who successively masters all the Islamic sciences. When the Dutch governor demands Datu Museng's wife Maipa Deapati as a concubine, they contemptuously refuse to obey. They declare their deep devotion to each other and vow to consummate a mystical union with one another in the grave. Significantly, however, their violent deaths are interpreted not as a form of martyrdom at the hands of an infidel, but as a freely chosen sacrifice at the hands of a fellow Muslim. Their self-sacrifice is the culmination of their lifelong quest to achieve mystical union with one another and with God. Like the tomb of Shaikh Yusuf and Daeng Nisanga, the joint tomb of Datu Museng and Maipa Deapati in Ujung Pandang remains a source of mystical blessing to this day.

The Bureaucratic Power of the VOC and the Predatory Slave Trade, 1667–1780

During the seventeenth century, the VOC was engaged in creating a radically new kind of power. It was a mercantilist corporation run by a rational bureaucracy of interchangeable officials whose explicit aim was to produce maximum profits for the shareholders back home. During the eighteenth century, the VOC acquired a growing number of territories and its officials were expected to administer a growing number of subjects. In South Sulawesi, the VOC became the overlord of Marusu' to the north of Ujung Pandang; Galesong and Polombangkeng to the south; Bantaeng, Bulukumba, and the ten little realms of the Bira peninsula to the east; the sultanates of Sumbawa, Tambora, and Bima across the Flores Sea on the island of Sumbawa; and the island of Selayar lying off the point of Bira.

Since VOC officials were rotated frequently throughout the East Indies, a means had to be found to pass on the knowledge they had accumulated about the complex marital and political relationships among the local elites, and of local laws and customs. A principal means of doing so were the *memorie van overgave*, memoranda of transfer, which summarized the chief political developments in an area for the benefit of the next official. The information contained in them had to be as explicit as possible, since their recipients were anonymous and might be entirely ignorant of the local situation.

In practice, official memoranda were never sufficiently detailed to provide a new official with all the knowledge he needed. The deficit was filled by the Indo-European offspring of Dutch officials and local women. A large community of these grew up around each Dutch outpost since the VOC did not allow women to sail past the Cape of Good Hope and most Dutch officials and soldiers contracted temporary liaisons with local concubines. VOC officials came to rely on these communities for interpreters and for knowledge of the local political situation. This was true of the VOC fortresses that were established in Bantaeng and Bulukumba in 1737. A careful reading of the *Regeerings Almanak* reveals that over the next century a number of Indo-European families came to dominate a range of lesser government offices in the area, such as interpreter, book keeper, and those in charge of "orphans," that is, the illegitimate offspring of European soldiers and merchants.

The slave trade was the principal source of wealth for both the VOC and its officers throughout the eighteenth century. The central role of Ujung Pandang in the regional slave trade dated back to the constant warfare between the VOC and Gowa that lasted from 1666 until 1677. These wars generated a large volume of captives who were sold as slaves. In 1676, 77 percent of the population of 1,400 in the residential quarter of Vlaardingen in Ujung Pandang fell into the category of slave, debt bondsman, or freed slave (*mardijker*). In the 1680s, ethnic Bugis and Makassar made up over 30 percent of all slaves in VOC possessions throughout Indonesia.

As the military situation stabilized during the eighteenth century, Dutch officials in charge of these territories became the primary suppliers of the slaves exported from Makassar. They assumed the power previously enjoyed by local chiefs to enslave people as punishment for certain crimes or in cases where a miscreant was unable to pay a fine (Nederburgh 1888). VOC officials in South Sulawesi typically supplemented their meager salaries by trading slaves on the side. In addition to their crucial functions as cultural brokers and interpreters, the descendents of VOC employees and local women also became specialists in the slave trade. In 1730, 71 percent of the total population of Ujung Pandang was enslaved. The average male mestizo in Makassar owned ten slaves, while the average Chinese and European males only owned five (Sutherland 1983: 268–269).

In 1754, Governor van Clootwijk ordered the compilation of a compendium of native laws for the use of VOC officials in charge of outlying territories. Sometime after his arrival in Makassar in 1848, Matthes collected three Makassar versions of these laws for the Netherlands Bible Society (manuscript numbers 25, 26, and 27). The Dutch version was published in a legal journal in 1853 (Brunsveld van Hulten 1853). Various translations of and commentaries on the Makassar version appeared in the 1880s (Matthes [1883] 1943: 36; Nederburgh 1888; Niemann 1889; see van den Brink, 1943: 56).

A principal objective of the compendium was the regulation of the slave trade. In his instructions to Willem Delfhout, the new resident of Bantaeng and Bulukumba, Governor van Clootwijk complained about the way the previous officials had abused their position to enrich themselves.

> All manner of robbery and extortion have crept in under the name of local custom, to the point that one has not scrupled to throw those who were impecunious in chains and sell them as slaves to satisfy the overweening love of money of the resident, experience having taught in this manner that such actions would serve as a means to corrupt the native chiefs, to twist the prescribed laws according to their sensualism and so as best to agree with their own interests, as became apparent in the case of several former residents, so it is found appropriate that a compendium be made of the laws, like those that were and still are in use by the Chiefs of Bonie and Goa, and which are faithfully recorded below. (van Clootwijk [1755] 1919: 150–151)

While it might appear from this passage that van Clootwijk was a high-minded reformer, his main concern was to clamp down on the diversion of profits from the VOC into the pockets of the residents.

> The VOC itself had trouble getting slaves from Makassar, until Governor Clootwijk (1752–1756) introduced a new system which allowed him to export about 1,500 "pieces" during his tenure. It seems probable that this was done by contracts with burghers, who would undertake to provide so many head until the demand from Batavia for, for example, 1500 men and 50 women for the artisans' quarter was finally filled. . . . If Clootwijk was proud of his 400 per annum, and the estimate of 3,000 is correct, then the private trade was more than six times that of the Company trade. (Sutherland 1983: 270)

The estimate of 3,000 slaves being exported from Makassar to Batavia each year throughout eighteenth century is "roughly equal to Dutch exports from West Africa, or Warren's figures for Sulu" (Sutherland 1983: 270). Combined with an estimate that each slave produced 100 guilders of profit, the slave trade alone brought in 300,000 guilders a year, a revenue stream that stopped when the Makassar-Batavia slave trade came to an end in 1820. After that, slaves continued to be exported from the Bugis ports of Bone and Pare Pare to Kalimantan.

In the eighteenth century, judicial fines "formed one of the main sources of income for the chiefs. Many residents of the Celebes in the 18th century, who did not wish to make do with their frugal salary, levied fines with their example in mind" (Rookmaker 1924: 515–516). The most common way for VOC officers in outlying areas to reduce people to slavery was to impose a fine so large that a person could not pay it. The imposition of *tunra*, a fine payable to the state, on top of *sapu'*, a compensation payment made to the family of the victim, was only in force in areas governed by Dutch residents. *Tunra* was not in use in areas governed by Gowa and Bone, as stated in Article 83: "Compensation (*sapu'*) and a fine (*tunra*) are imposed for murder and theft only in Bulukumba, Bantaeng and Marusu', while in Ujung Pandang, Bontoala and Gowa [*tunra*] is not imposed, but only the customary compensation." The compendium thus codified the practice so strongly criticized in the cover letter: the ability of VOC officers to sell their subjects into slavery to satisfy their own greed.

A Konjo Makassar translation of the eighty-eight articles contained in the compendium was preserved in Bira at least until 1936, when a copy was made for *Controleur* de Roock. The eighty-eight articles in the Bira manuscript may be grouped under the following headings:

1. Legal definitions and procedures (eleven).
2. Criminal matters (thirty-one): including theft (eight), crimes of violence (thirteen), and sexual transgressions (ten). In all cases, punishments are graded according to the rank of both the perpetrator and the victim, with nobles having almost complete freedom to act as they like toward their own slaves.
3. Civil matters (thirty-three): including marriage and inheritance (eight), compensation for civil damages (thirteen), and debts (twelve). All marriage payments are graded according to the relative ranks of the groom and the bride.
4. The regulation of rights in persons, including those of lords over subjects (*jannang*) and of masters over slaves (*ata*) and pawns (*tunitaggalaka*) (fourteen).

The articles in the fourth category evoke a world in which freedom could be lost at any moment through an inability to pay one's debts or by being subjected to arbitrary fines. Relations between the powerful and the powerless were governed by a harsh mercantile code in which there was a lively commerce in human beings, especially women and children. It was also a world in which intimacy between masters and slaves was common and in which many children of mixed Dutch and Indonesian parentage were produced.

The compendium served as a guide for Makassar chiefs in Dutch-administered territories well into the nineteenth century. As late as 1884, Engelhard reported that its provisions had "penetrated into the native household in such a manner, that in many respects they still serve at present as a guide-book for the chiefs' own judgments." He felt that without an understanding of its contents, "many of the native usages, customs and habits must appear mysterious" (Engelhard 1884b: 828).

Governor van Clootwijk's successor, Roelof Blok (r. 1756–1760) carried van Clootwijk's rationalization of the VOC bureaucracy still further. In 1758, he implemented the code of criminal law written by van Mossel in 1736 for the Coromandel coast of India and the code of administrative law written by Taillefert in 1755 for Bengal (van Kan 1935). Blok also took an active interest in local customs and history. He composed a history of the island based on local chronicles. It shows much of the same appreciation of the relevance of local ethnographic detail to sound governance that was later displayed by the British governor of Java, Sir Stamford Raffles, who had access to his papers (Blok [1759] 1848; von Stubenvoll 1817; Raffles 1830).

Due to the efforts of van Clootwijk and Blok, Cornelis Sinkelaar (r. 1760–1767) was the first governor to take office with a full set of administrative, criminal, and customary laws at his disposal. The fact that the VOC governor was now regarded more as a territorial lord than as the agent of a trading corporation was clearly illustrated by the role Sinkelaar was expected to play in a complex succession dispute that took place in the kingdom of Sumbawa. The fact that he did not have the local legitimacy to play this role effectively was illustrated equally clearly by the tragic outcome.

Dynastic Machinations in Gowa, Tallo', and Bone, 1677–1762

When Arung Palakka took possession of Gowa's final stronghold at Kale Gowa in 1677, Gowa's ability to defy the VOC militarily was at an end. Tallo' did not have much in the way of military power in South Sulawesi either, but it did have a far-flung network of political and marital alliances scattered around eastern Indonesia. The Hapsburg-like skill with which Tallo' placed its descendents on thrones throughout eastern Indonesia was a continual source irritation to Bone and the VOC during the eighteenth century. Governor Adriaan Hendrik Smout (r. 1737–1744) put the matter bluntly as follows: "The Maccassaar have done as well since the loss of their power by the coupling of their princesses as they did before by their use of the sword" (Ligtvoet [1875] 1987: 11).

Tallo's ability to manipulate lines of succession in this way derived from the fact that noble rank and rights to high office were transmitted bilaterally. In theory, the child of a ruler's highest-ranking spouse had the best claim to the throne. In Bugis kingdoms, a woman was often selected as the ruler if she outranked all her brothers. In Makassar kingdoms, the selection of a female ruler was extremely rare. Women still played an important role in the selection process, however, because it was often the rank of a man's mother that placed him ahead of his half brothers.

There was much scope for manipulation in this system since most powerful rulers married many times and could produce large numbers of potential heirs. Princes could marry several wives at once, but even princesses often married several husbands in succession. Rulers sought the highest-ranking marriage partners for their children, and often arranged marriages with the children of rival rulers. Over time, the different royal houses of South Sulawesi and neighboring islands such as Kalimantan and Sumbawa were knit into a single network of kinship and marriage.

The most important source on Tallo's marriage politics in the eighteenth century is *The Diary of the Kings of Gowa and Tallo'*. The first part of this remarkable text appears to have been composed between 1713 and 1731 by Shafi al-Din some years before he became sultan of Tallo' (r. 1735–1760). A copy of the diary was made in 1730 and then continued until 1751, probably by Shafi al-Din's half brother, Zain al-Din, many years before he became sultan of Gowa (r. 1769–1777) (Ligtvoet 1880: 1–3). The diary provides a detailed account of every marriage, divorce, birth, and death of political significance to Tallo' for most of the seventeenth century and for the first half of the eighteenth century. It documents the tactics used by of the royal house of Tallo' to maintain its position at the top of the regional marriage system even after the fall of Gowa. As a kinship-based political text produced by Makassar nobles, it provides a revealing contrast to the bureaucratic *memorie van overgave* produced by VOC officials. Figure 4.1 is drawn largely from this source, supplemented by a number of secondary Dutch sources (Buddingh 1843: 443–450; Zollinger 1850; van Hoevell 1854: 158–161; Ligtvoet [1875] 1987).

Arung Palakka's initial attempts to consolidate its power over South Sulawesi were hampered by the fact that he was unable to father offspring of his own. He pursued his long-term ambitions for the royal house of Bone by arranging a series of marriages for his nephew and heir apparent, La Patau, who was known after his installation as Sultan Idris (r. 1696–1714). In 1686, Arung Palakka arranged for Idris to marry We Umung, the daughter of Sultan Muhammad of Luwu' (r. 1662–1704). This marriage produced a daughter, Batari Gowa, who was to dominate the politics of Bone for the first half of the eighteenth century. In 1687, Arung Palakka arranged for Idris to marry Mariam, the daughter of Sultan Abd al-Jalil of Gowa (r. 1677–1709). This marriage produced a son, Ismail, who was to rule both Gowa and Bone, in succession.

When Arung Palakka died in 1696, his successor, Sultan Idris, took over the task of consolidating Bone's marriage alliances. In 1702, Idris arranged for Ismail to marry the daughter of Sultan Abd al-Qadir of Tallo' (r. 1670–1709). This marriage produced a daughter, Amira Arung Palakka, who was to play a central role in the politics of South Sulawesi from the time of her marriage in 1735 until her death in 1779.

Sultan Abd al-Qadir of Tallo' died in 1709 and was succeeded by his son, Siraj al-Din. Sultan Abd al-Jalil of Gowa died the next month without having fathered a son. Sultan Idris of Bone took advantage of the fluid situation in Tallo' and Gowa to arrange for his own son to be selected as Sultan Ismail of Gowa. Had Ismail kept control of Gowa, it might have been permanently absorbed into the kingdom of Bone. But by 1712, Sultan Siraj al-Din of Tallo' had grown strong enough to persuade the Nine Electors of Gowa to depose Ismail and to install him in Ismail's place.

Sultan Idris was succeeded as ruler of Bone by Batari Toja, his daughter by Princess We Umung of Luwu'. Batari Gowa abdicated the following year in favor of her half brother Sulaiman. Like Ismail, Sulaiman was a son of Princess Mariam of Gowa and so was also in a position to try to unite the kingdoms of Bone and Gowa. When Batari Toja's mother was deposed as Queen of Luwu' in 1719, Batari Toja was selected to take her place. This changed the balance of power and in 1720, Batari Toja was able to persuade Bone's Council of Electors to depose Sulaiman and restore her to the throne of Bone, thus briefly uniting the kingdoms of Luwu' and Bone under a single ruler.

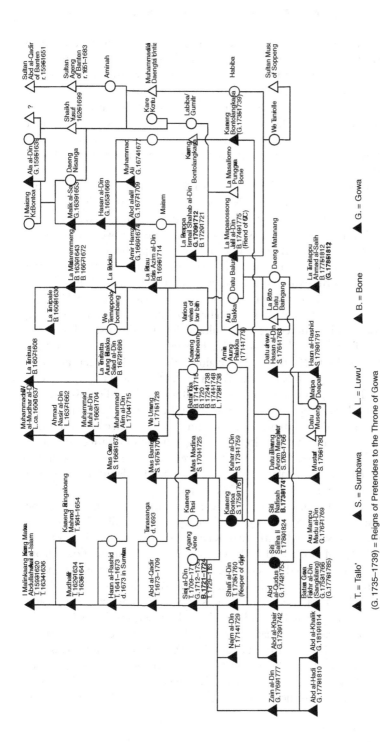

Figure 4.1 Royal Alliances, 1654–1812

▲ T. = Tallo'   ▲ S. = Sumbawa   ▲ L. = Luwu'   ▲ B. = Bone   ▲ G. = Gowa

(G. 1735–1739) = Reigns of Pretenders to the Throne of Gowa

The royal house of Bone had not given up on trying to absorb Gowa, however. Soon after she was reinstalled as ruler of Bone in 1720, Batari Gowa abdicated again in favor of her brother Ismail, who had been deposed as sultan of Gowa in 1712. Ismail proved to be just as ineffectual as ruler of Bone as he had been as ruler of Gowa, however. Sultan Siraj al-Din of Gowa and Tallo' was able to depose him as ruler of Bone in 1721 and have himself installed as the sultan of Bone, uniting the three kingdoms under one ruler for the first time (Buyers 2000–2005).

Siraj al-Din abandoned his claim to Bone in 1724 and the Electors of Bone initially chose Ismail's brother Mansur to replace him. After only four days, however, they decided to reinstall Batari Toja. In 1728, she was installed as the Datu of Soppeng as well, uniting the three most powerful Bugis realms of Luwu', Soppeng and Bone under a single ruler. Batari Toja's next move was to try to gain control of the territories of Bulukumba and Bantaeng that lay to the south of Bone. These territories had been transferred from Gowa to the VOC in 1667 by the Treaty of Bungaya. The VOC had given them to Arung Palakka and his immediate heir, Sultan Idris, to administer. Upon the death of Idris in 1714, the VOC had reasserted its control of the territories. In 1730, Batari Toja claimed that they had been given by the VOC to Bone in perpetuity. When the VOC refused to return them, Batari Toja abandoned her residence at Bontoala' near Fort Rotterdam and withdrew to her palace at Cenrana in the heartland of Bone.

Batari Toja's fortunes now took a turn for the worse as a result of a complex alliance that developed among a group of disaffected nobles from Gowa, Bone, and Wajo'. They were led by a Gowanese noble called Karaeng Bontolangkasa, who married the daughter of Sultan Amas Madina of Sumbawa in 1723. When the sultan died in 1731, his widow forced their daughter to divorce Bontolangkasa and marry the new Sultan of Sumbawa, Qahar al-Din (r. 1731–1759). Humiliated, Bontolangkasa allied with Aru Kayu, a former husband of Queen Batari Toja of Bone and with Aru Sinkang, a rebellious noble from Wajo'. They resolved to seize the thrones of Gowa and Bone and to drive the Dutch out of South Sulawesi (Buddingh 1843: 443–444).

Bontolangkasa used Sumbawa as a base to organize an uprising against both Siraj al-Din and the VOC. In 1734, his forces landed in South Sulawesi and routed the army of Gowa in battle. In November, 1735, Siraj al-Din was forced to abdicate the throne of Tallo' to his son, Shafi al-Din (r. 1735–1760), and the throne of Gowa to his grandson, Abd al-Khair (r. 1735–1742). In 1736, Queen Batari Toja of Bone was forced to flee from her palace in Cenrana and to seek the protection of the VOC in Ujung Pandang. In 1738, the forces of Aru Kayu installed Siti Nafisah on the throne of Bone. She was the daughter of Amira Arung Palakka by her husband Sultan Shafi al-Din of Tallo'. This marked the beginning of Amira's centrality to the politics of South Sulawesi.

In 1737, Arung Singkang landed in Wajo' and took control of the capital at Tosora. He raised an army in Wajo', Soppeng, and Bone and marched south to join Bontolangkasa in an attack on Fort Rotterdam. Bugis forces loyal to the VOC and to the deposed queen of Bone, Batari Toja, rallied around the Dutch under the command of Batari Gowa's half brother, La Mapasossong, the son of Sultan Idris by a low-ranking wife. When Bontolangkasa and Arung Singkang laid siege to Fort Rotterdam itself in 1739, La Mapasossong played a key role in defeating them. Bontolangkasa died from the wounds he received in this battle. La Mapasossong went on to help the Dutch drive the hostile forces out of Bone and Soppeng in 1740 (Le Roux 1930: 212–213). When Siti Nafisah died in 1741, Batari Toja was installed for the fourth time as ruler of Bone. The attempt to subdue Wajo' failed in 1741, and that kingdom remained relatively free of Dutch influence until the end of the nineteenth century (Noorduyn 1972).

Batari Toja died in 1749, and La Mapasossong was chosen to succeed her at the insistence of the VOC. In his memorandum of transfer written in 1756, Governor van Clootwijk wrote of the friendship he had enjoyed with La Mapasossong since his arrival in Makassar twelve years previously. He would not ordinarily have been considered for the throne because of the low rank of his mother, and his installation evoked strong resistance from the Electors of Bone. Van Clootwijk only managed to persuade the Electors to formally install him as Sultan Jalil al-Din of Bone in 1752 (Le Roux 1930: 227).

The War of the Sumbawan Succession, 1760–1762

Sultan Qahar al-Din of Sumbawa died in 1758, and his wife, Karaeng Bontoa, was installed as the sultana. In 1762, Sumbawa's council of five electors deposed her and installed the Datu of Jarewe as Sultan Hasan al-Din (r. 1762–1763). The chief of the council, who bore the title of Nene Rangan, forced his daughter to divorce a noble called Mille Ropia in order to marry the oldest son of the new sultan. Humiliated by the loss of his wife, Mille Ropia entered into an alliance with the Datu of Taliwang (later known as Sultan Jalal al-Din) and with some Bugis refugees from Wajo'. They resolved to replace both the Nene Rangan and the sultan with candidates from their own group.

Hasan al-Din responded to this threat by inviting Balinese forces to cross from the neighboring island of Lombok and to intervene on his side. Their leader, Gusti Ngurah, agreed to do so. He had a grievance of his own against Mille Ropia and Datu Taliwang, since they had recently abducted one of his wives during a raid on Banjarmasin in Kalimantan.

To counter this Balinese threat, Datu Taliwang enlisted the support of the VOC resident in Bima, Johann Tinne (1758–1764) by promising him a gift of one hundred slaves. Tinne persuaded Governor Sinkelaar to send troops to help Datu Taliwang. In November, 1763, three sloops carrying twenty-one European soldiers arrived under the command of the resident in Selayar, Jakob Bikkes Bakker (r. 1758–1764; see Sutherland 1983: 270). Mille Ropia was killed in the subsequent battle and the Dutch succeeded in driving out the Balinese. They captured Sultan Hasan al-Din and brought him back to Ujung Pandang in February, 1764 (Noorduyn 1987b). Datu Taliwang was duly installed as the Sultan of Sumbawa.

There matters might have rested, except that Datu Taliwang's Dutch ally, Tinne, died soon after, on June 25, 1764. The commander of the Dutch expedition to Sumbawa, Bakker, took Tinne's place as Resident in Bima (1764–1768). Fearing that the death of Tinne would undermine his influence with Governor Sinkelaar, Datu Taliwang sent a mission to Ujung Pandang at the end of 1764. It included the current Nene Rangan, another of the five Electors, and the Datu of Busing. The Datu of Busing was also the chief of Re, one of ten districts in Sumbawa proper, and the governor of one of the four districts into which the capital was divided.

They arrived too late: Sinkelaar had indeed decided that he had been misled by Tinne about Datu Taliwang's right to the throne of Sumbawa. On February 9, 1765, Sinkelaar signed a treaty recognizing the Datu of Jarewe, Hasan al-Din, as the true sultan of Sumbawa. Sinkelaar ordered the arrest of the three emissaries of Datu Taliwang to prevent them from returning to Sumbawa with the news. Two of the emissaries surrendered, but on March 4, 1765, Datu Busing "ran amok" in Ujung Pandang. Governor Sinkelaar wrote that "after noon two of the so-called delegates were brought to me while the third, escaping, went to his dwelling and took up a position there with his people," where he was finally killed (Noorduyn 1987b: 34–35).

For the Dutch, the dispute over the throne of Sumbawa was only one example of a long series of similar disputes in which they were involved. For the Makassar subjects of the VOC, it came to embody everything they hated about being subject to infidel overlords. The Datu Busing who ran amok in 1765 became the hero of an epic poem, the *Sinrili' Datu Museng* (see map 4.1).

The Sinrili' Datu Museng, 1850–1988

Sinrili' are a popular Makassar narrative genre that usually have a meter of one foot of five or six syllables alternating with another foot of eight syllables (Matthes [1883] 1943: 326). The subject matter of *sinrili'* usually refers to events that took place in the time of the "*Kompeni*," as the VOC was known. A point made by Ileto about the popular narratives that developed in the Christian Philippines holds equally true for Makassar *sinrili'*.

Map 4.1   The World of Datu Museng

One characteristic of such Tagalog sources as narrative poems and songs is their apparent disregard for accurate description of past events. But factual errors, especially when a pattern in their appearance in discerned, can be a blessing in disguise. . . . When errors proliferate in a patterned manner, when rumors spread "like wildfire," when sources are biased in a consistent way, we are in fact offered the opportunity to study the workings of the popular mind. (Ileto 1979: 14)

It is also true that the line between written and oral versions of *sinrili'* was crossed as frequently as it was in the case of Tagalog narratives (Ileto 1979: 15).

The *Sinrili' Datu Museng* first came to the attention of Europeans in the 1850s when Dutch missionaries recorded two different versions of it. One of these missionaries, Benjamin Matthes, had been commissioned by the Netherlands Bible Society to translate the Bible into Makassar and Bugis. Matthes collected a large number of local manuscripts written in the *lontara* script. He also hired local scribes to record various pieces of oral literature, especially the short poems called *kelong* and the epics called *sinrili'*. The *Sinrili' Datu Museng* recorded by Matthes in 1852 contains about 1,150 lines of thirteen or fourteen syllables each.

The Sinrili' Datu Museng (Matthes, Gowa, ca.1852)

Datu Museng is a loyal servant of Datu Taliwang, the sultan of Sumbawa. The sultan learns that the Datu of Jarewe is in Ujung Pandang and is on the verge of persuading the governor to recognize him as the legitimate Sultan of Sumbawa. He sends Datu Museng to plead his case. Datu Museng brings his wife Maipa along. When the governor learns of her beauty, he demands her as a concubine, offering to exchange forty slave women for her. Datu Museng refuses. Together, he and Maipa decide it would be better to die and reunite in heaven than to submit to an infidel. Maipa washes, prays and allows Datu Museng to cut her throat. Datu Museng discards his protective amulet and fights to the death. The Karaeng of Galesong kills him with a lance and brings his head back to the governor. (condensed from Matthes 1860: 529–563)

The moral contrast between the pious Datu Museng and the lascivious governor is made clear in the series of epithets that are used each time their names are mentioned.

Karaeng I Datu Museng, who is firm in faith, generous in alms-giving to those who chew betel and to the poor; who pities the unfortunate; / who turns from the forbidden and avoids the ill-advised; who is lit by an inner light; / who is fortunate in deed and pure in thought; whose will is not thwarted, nor are his undertakings; / who is the descendent of prophets, the commander of the faithful; / who is the offspring of spiritual masters (*guru*) and the son of the wise (*panrita*).

In contrast to Datu Museng's sterling moral qualities, the governor is identified chiefly in terms of his bizarre physical features and peculiar cultural practices.

The Great Lord, the world-mighty, the world-ornament; / who draws a long dagger to strike those who kneel; / who wears a broad hat, whose hair is yellow, whose skin is white; / whose teeth are unfiled, who is uncircumcised.

Matthes explained the cause of the events recounted in the *Sinrili'* solely in terms of the succession dispute that took place in Sumbawa in the 1760s. He refused to believe that a Dutch governor's lust could have played any role in precipitating the final tragedy (Matthes 1860: 511–514). In view of the scale of the extensive slave trade overseen by VOC governors such as van Clootwijk and Sinkelaar, however, Matthes's Victorian incredulity about an eighteenth-century Dutch official's willingness to use force to satisfy his "foul lust" seems anachronistic, to say the least.

The second version of the *Sinrili'* was published by William Donselaar, a missionary stationed in Bantaeng to care for the Indo-European community that had grown up around the Dutch fort there.

The Sinrili' Datu Museng (Donselaar, Bantaeng, ca. 1852)

Datu Museng elopes with Maipa Deapati, the daughter of the Datu of Jarewe, a vassal of the sultan of Sumbawa, who is himself a vassal of the Great Lord, the Dutch governor of Ujung Pandang. Datu Jarewe goes to complain about Datu Museng to the governor. Datu Museng follows to defend himself, but the governor falls in love with Maipa and tries to seize her by force. Maipa asks Datu Museng to kill her first. She bathes in consecrated water and offers her throat to his knife. He kills her, wraps her in a shroud and recites the appropriate prayers from the Koran. Then he goes out and fights until he is exhausted. He cannot be shot or stabbed because he is invulnerable. The Karaeng of Galesong finally kills him by striking him in the head with the butt of a rifle. (condensed from Donselaar 1855)

In this version, the story is motivated as much by contradictions internal to Makassar society as it is by the uncontrolled power and appetites of the Dutch overlords. As we will see, the detail about the elopement of Datu Museng with the daughter of a nobleman of higher rank is central to the version I recorded in Ara in 1988.

The Sinrili' Datu Museng in Selayar, Bira, and Ara

In 1988, I recorded two slightly different versions of the *Sinrili' Datu Museng* in Ara, one by Dessibaji' Daeng Puga and one by Muhammad Nasir. Both versions were derived from Haji Abdul Hae' of Bira. Haji Hae was the son of Pu Ijon, a nobleman from the island of Selayar. Pu Ijon's sister, Karaeng Tawa, was married to the regent of Bira, Baso Daeng Raja (1849–1884), who was in office at the time that Matthes and Donselaar collected their versions of the *Sinrili'*. Haji Hae' was thus a first cousin of Baso Daeng Raja's son and successor, Andi' Mulia Daeng Raja (r. 1901–1914, 1931–1942).

During the 1920s and 1930s, Haji Hae' taught the *Sinrili'* to his own son, Haliki, and to Muhammad Nasir and Daeng Pagalla of Ara. Daeng Pagalla served briefly as *gallarrang* of Ara in 1913–1915, but was removed from office in favor of a commoner called Gama Daeng Samana (see chapter 6). Daeng Pagalla spent the remaining thirty years of his life as a simple farmer and mystic. He inherited a library of Islamic manuscripts that had belonged to the *kalis* of Ara, some of which had been passed down from the founder of the saintly lineage of Bira, Haji Ahmad the Bugis (see chapter 3). Daeng Pagalla had no children of his own, but his sister's grandson, Dessibaji, was his constant companion. He taught him the *Sinrili'* while they worked together in the fields, and it was Dessibaji's version that I first recorded.

Haji Hae's version contains many events from the early part of Datu Museng's life that are not mentioned in the versions recorded in the 1850s. These include his early years as a poor orphan, his acquisition of the rudiments of Islamic knowledge in the palace of the sultan, his rejection by the sultan as a suitor for his daughter, his acquisition of esoteric knowledge in Mecca and Medina, and his triumphant elopement with Maipa Deapati upon his return. The way these earlier episodes in Datu Museng's life are treated transform the *Sinrili'* from a simple narrative of heroic resistance to the VOC into an elaborate Sufi allegory about separation from and reunion with the divine Beloved.

Daeng Pagalla, Dessibaji, and Muhammad Nasir all studied Sufism in the village of Cikoang. Hamonic gives an account of what was taught in Cikoang in the early 1980s.

> Traditionally, we are told, religious education here comprised four degrees: *Angngaji*, apprenticeship in reading and writing the Qur'an; *Assarapa'*, apprenticeship and knowledge of Arabic grammar; *Assarea'*, apprenticeship in Islamic law (*shari'at*) in Arabic and in the Makassar language written in Arabic characters; and finally *Attareka'*, apprenticeship in gnosis (*ma'rifat*) and in spiritual truth (*hakikat*). This last degree, which develops the discussion of religious problems relating to the creation of the world and to the figure of the Prophet, presupposes of course the perfect mastery of the degrees that precede it. It thus requires a secret initiation that takes place by word of mouth from master to disciple. (Hamonic 1985: 179–180)

According to Hamonic, many of the manuscripts left by the founder of Sufism in Cikoang, Jalal al-Din al-Aidid, were destroyed during the Darul Islam insurrection in the 1950s. Many other works were still in use in the 1980s, however, including the *Sharab ul-'ashiqin* of Hamzah Fansuri and several works by Nur al-Din al Raniri (Hamonic 1985: 180).

While Dessibaji was a traditionalist in religious matters, Nasir was a staunch modernist. This explains why Nasir's version of the *Sinrili'* differed from Dessibaji's on certain points having to do with differences between traditionalist and modernist Islamic doctrines during the twentieth century. While Muhammad Nasir was open to the orthodox teachings of the Khalwatiyya, he rejected the antinomian claims of some mystics in Cikoang that they had transcended the necessity to obey the *shariah* law. He explained his view of the relationship between the *shariah* and *tariqa'* with a parable about sailing. The *shariah* is like a boat and the *tariqa'* is like the crew. Without a material vessel, you cannot go anywhere. Hence religion requires that there be material bodies that may be moved in accordance with God's explicit commands. But a boat without a crew will drift aimlessly. Blindly following the *shariah* without knowing its inner meaning is just as pointless as trying to sail without a boat.

Despite such differences of detail, the overall structure of the two versions is the same, and I summarize them together. I have divided the story into four parts based on Datu Museng's journeys to and from four places: his home in Taliwang; the center of traditional royal authority in Sumbawa Lompo; the center of charismatic Islamic authority in Mecca; and the center of bureaucratic VOC power in

Ujung Pandang. I argue later that these four divisions correspond to the four stages of the Sufi path: *shariah*, *tariqa'*, *haqiqa*, and *marifa*. The version recorded by Dessibaji in Ara in October, 1988, is twice as long as the one recorded by Matthes in the early 1850s, containing about 2,000 lines of text. The recording runs for three-and-a-half hours. The version of Muhammad Nasir is more than half as long again, containing about 3,500 lines of text. The recording runs for four-and-a-half hours. The meter is much looser in the Ara versions than in the one recorded by Matthes. Most phrases contain about eight syllables and are paired with another phrase of eight syllables. Sometimes, a shorter phrase of six syllables is followed by two shorter phrases of similar length. Thus each set of phrases is about one third longer than a line in the Matthes version.

The Sinrili' Datu Museng (Dessibaji' and Muhammad Nasir, Ara, 1988)

1. Taliwang to Sumbawa Lompo [Dessibaji' lines 1–250; Nasir lines 1–45]

Datu Museng is known in his youth by the name Baso Mallarangang, "The Forbidden One." He is raised in the remote village of Taliwang by Nene' Rangan, a mysterious being who was delivered in a cave by a tiger shaman and brought up by a snake. When Baso Mallarangang reaches adolescence, Nene' Rangan allows him to study the Koran with the Imam of Sumbawa Lompo. In just three days he memorizes and understands not only the Qur'an, but also the poetical Life of the Prophet written by Jaffar al-Barzanji (d. 1766). [Reference to this text is omitted in Nasir's "modernist Islamic" version for reasons explained in chapter 7.]

Also studying with the imam is Maipa Deapati, the young daughter of the Karaeng and the most beautiful woman in the kingdom. She has learned nothing after three months, and the Imam tells Baso Mallarangang to tutor her. Impatient with her at pointing to the wrong letters while he is reciting, he slaps her hand, causing her heirloom ring to fall through the bamboo slats of the floor. Baso Mallarangang retrieves it from under the house, but vows to return it only after he has married her. After this, whenever Baso refers to his love for Maipa, he claims that it has existed "since they were in the womb." Informants interpreted this to me that they were actually opposite-sex twins who were separated at birth.

Baso Mallarangang returns to Taliwang and sends a messenger to the King asking for Maipa's hand in marriage. The messenger is rudely rebuffed, since Datu Museng belongs to the commoner class of *gallarrangs*, and noble women cannot marry beneath them. He is told that Datu Museng should keep his place, since "his father was only a vassal lord, his mother of *gallarrang* rank: dogs stay beneath the house, while slaves look after the buffalo; slaves look after the horses, while cats guard the hearth."

2. Taliwang to Mecca [Dessibaji lines 250–1400; Nasir lines 46–1974]

Humiliated, Baso Mallarangang sells his buffalo to pay for a pilgrimage to Mecca. He studies the esoteric knowledge for seven years in Mecca and seven years in Medina. This knowledge gives him mastery of the arts of love and war. At the end of his stay, he performs the rituals of the *hajj* and takes a new, noble name, Datu Museng. He returns to Taliwang on the eve of Maipa Deapati's arranged marriage to her first cousin, Dammung Alasa of Lombok. Datu Museng becomes so depressed he sleeps all the time and forgets to pray. Nene' Rangan finally, persuades him to attend the wedding. [In Nasir's version, Datu Museng now recites the *dhikr kasambandia* and the creed of the Khalwatiyya Sufi order that banishes all fear, and performs the *beja'beja'na* ritual of Madina, which gives invulnerability.] As they approach the festivities, Datu Museng warns Nene' Rangan not to react when he deliberately provokes the contempt of the crowd: it is a ruse to draw Maipa to the window so he can bespell her.

Datu Museng and his grandfather show up just in time to disrupt the proceedings. Datu Museng challenges the Sumbawans to a contest playing raga ball. In order to get Maipa Deapati to look out the window, he plays the fool until the entire crowd is laughing at him. Then he exercises his magical powers on the ball, making it stay up in the air as long as he wishes. Maipa finally bathes and dresses and sits against the central post of the house before going to the window. Datu Museng catches Maipa looking out the window and casts a spell on her, causing her to faint. He continues outplaying his rival, Dammung Alasa, mocking his manhood. The Dammung attacks him and soon Datu Museng and Nene' Rangan are fighting the entire army of Sumbawa. Due to their invulnerability magic, they put them all to flight.

Before leaving, Datu Museng makes the raga ball descend into Maipa's house, turn into a lizard, and enter her belly. Maipa falls grievously ill and none of the healers can help her. Finally, the ruler of the Malays tells her father that only Datu Museng can cure her. [In Nasir's version, when Datu Museng is summoned he throws the king's words back at him, saying that he is unworthy to enter the house. After many entreaties, he is persuaded to go.] When he arrives at the palace of Sumbawa, he enters Maipa's bedchamber, cures her with ordinary water, and shows her the lost ring, vowing to return it only after they are married. He goes away, but a week later at midnight casts a spell on her, causing her to awaken and to insist on going down to the well to fulfill a vow. Datu Museng causes her escort to flee by sending down a rain of ashes, and Maipa makes her way to his house. It is dark, but when she enters and sits against the central post it is filled with light. They are married by the Guru Keramat of Taliwang. Datu Museng returns her ring and they consummate their union. [In Nasir's version, Maipa warns Datu Museng when they enter the bedchamber that the wedding cakes from her father will be cannon balls. He tells her not to worry, to join him in reciting the *shahabat* and *dhikr*, to trust him as he is the student of the four Imams of Mecca and the Shaikh of Medina.]

3. Taliwang to Ujung Pandang [Dessibaji lines 1400–1750; Nasir lines 1975–3000]

When her absence is discovered, the Karaeng Dea Rangan sends his army after them, but Datu Museng and Nene' Rangan easily defeat it. Knowing now that Datu Museng's magic is too strong for him, the Karaeng hatches a scheme. He forges a letter from the Dutch Governor of Makassar, known as *Tuan Malompoa*, The Great Lord, ordering Datu Museng to come to Ujung Pandang and promising to appoint him as his successor. Maipa is fearful it is a trap, but Datu Museng reassures her that his invulnerability magic is such that no blade or bullet can pierce his skin. Datu Museng and Maipa Deapati set out on his great black ship, the I Lolo Gading.

[The crossing to Ujung Pandang takes up almost one-third of Nasir's version. It also contains much sailing lore not found in Dessibaji's version. In Nasir's version, the king of Sumbawa, Dea Rangan, writes a letter to the governor and encloses Maipa's portrait. Captain Laterre takes it on his ship Mangkinnaya to Ujung Pandang, along with Maipa's jilted fiancé, Dammung Alasa. The governor is immediately smitten by the portrait and vows to take the ring from her finger and her body from the lap of Datu Museng. Dammung Alasa comments that Datu Museng has already stolen Maipa from him. Their mission accomplished, they return to Sumbawa. Dea Rangan summons his spies from Lombok. He tells them to go invite Datu Museng to his house, pretending he is ready for a reconciliation because of the longing of Maipa's mother for her child. Datu Museng, Nene' Rangan, Maipa and their retinue set out for Sumbawa Besar, but Maipa stops outside the walls and says she cannot continue in to meet her father and mother, but will return to Taliwang and worry about the letter being sent to Ujung Pandang. Datu Museng continues on and is warmly received by Dea Rangan, who tells him he has received a letter from the governor asking Datu Museng to come. Dea Rangan says he can live in his palace in Galesong, which is plentifully supplied with rice, fish, and money. Datu Museng replies that when he was in Mecca he had received a sign that he was destined to mix with the earth in Ujung Pandang. Dea Rangan sends for the shipbuilders and carpenters of Ara to build him a boat. When felling the first tree for the boat, the ritual expert sees a sign that it will make only one trip, and will then be destroyed. Datu Museng says it does not matter and returns to Taliwang to collect Maipa. Datu Museng steers the boat across the Java Sea by magic, surviving pirate attack and storms. During the passage, the two of them discuss their coming death and prepare themselves for martyrdom and reunion in paradise.]

4. Ujung Pandang to Heaven [Dessibaji lines 1750–2025; Nasir lines 3000–3500]

They arrive in Galesong where the ruler is his kinsman and settle into a palace. When the governor sees Maipa sitting in a window, he becomes infatuated with her and sends the Public Prosecutor to offer Datu Museng forty concubines in exchange for her. Datu Museng indignantly refuses, and Maipa says she would rather die. The governor sends his soldiers, but he and Nene' Rangan run amok and put them to flight. [In Dessibaji's version, when Maipa realizes the next morning that she is the cause of scores of deaths, she says she would rather die than cause any more deaths or become the wife of a Dutchman. Datu Museng cuts her throat and arranges her on a chair as if she were still alive. He goes down to the sea and throws his amulets and keris into the sea, thereby making himself vulnerable to weapons. He asks his relative, the Karaeng of Galesong, to whip him to death, since his skin can still not be pierced by a blade. When the governor hears he is dead, he goes to the house and embraces Maipa. When he realizes she is dead, he jumps back, hits his head on a post and dies.]

[In Nasir's version, when Daeng Jarre reports his failure, the governor orders that Dutch, Javanese, and Madurese troops be assembled and sent to Kampong Galesong. The house is attacked, and the battle moves to the field of Karebosi. Nene' Rangan chases the troops back to the governor's house. The governor flees and hides in the prison. Maipa sits before Datu Museng and asks to be killed. Datu Museng does so and returns with Nene' Rangan to fight in Karebosi. Datu Museng duels with Raja Galesong and is finally shot by his bullet. News of these events reaches the governor general, who orders that the governor be brought before the High Court for trial. He is sentenced to be hung until dead.]

The Captain of the Javanese in Ujung Pandang buries Datu Museng by the shore where he fell, but the grave begins to move. After seven nights it lies next to that of Maipa Deapati, so that just as their spirits are reunited in the afterlife, their bodies are reunited on earth. Due to all the knowledge Datu Museng accumulated in Mecca and Medina, these twin graves become a sacred place where devotees implore his intercession with God.

Analysis and Comparison of the Four Versions

The *Sinrili' Datu Museng* grew greatly in length and complexity between 1850 and 1988. All four versions link Datu Museng to the Datu of Taliwang and cast the Datu of Jarewe as his enemy. Only

Matthes's version retains a reference to the original dispute between Datu Taliwang and Datu Jarewe over succession to the office of sultan. The other three attribute Datu Museng's presence in Ujung Pandang to the anger of Datu Jarewe over the abduction of his daughter. In all four versions, Datu Museng kills Maipa and allows himself to be slain by the ruler of Galesong to avoid dishonor at the hands of the Christian ruler. The greatest development occurred in the first half of the *Sinrili'*, when Datu Museng learns the Koran, is rejected by Maipa's father, acquires esoteric knowledge in Mecca, and returns to abduct her. These episodes acquire a status equal to the later episodes set on the boat and in Ujung Pandang.

The *Sinrili' Datu Museng* operates on at least three levels of meaning. On the first level of meaning, it reflects the distinctive culture of resistance that developed in the lands that were directly ruled by the VOC during the eighteenth century. Local political overlords were expected to officiate at Islamic rituals in their capacity as Defenders of the Faith and to form hypergamous marriage alliances with their peers and subjects in their capacity as successors to the *tomanurung*. Calvinist officials could do neither. In local eyes, Dutch power rested entirely on their superior military force, which was used primarily to advance their individual greed for wealth. South Sulawesi's primary export under VOC rule came to slaves who were either captured in battle or enslaved in lieu of payment of fines assessed by corrupt office holders. The governor and his subordinates are portrayed as completely immoral tyrants whose power rests entirely on military force. A radical separation between the local social hierarchy, the regional political order, and the global religious order thus arose.

The kind of martyrdom portrayed in the Sinrili' Datu Museng provided a way out for the most desperate and marginalized individuals elsewhere in colonial Southeast Asia as well. The practice was known as *parrang sabil*, "war in the path of God" among the Tausug of the southern Philippines.

> Group resistance against the Spanish prior to the middle of the 19th century was relatively well organized through the institutions of the Sultanate. But with the Spanish conquest of the town of Jolo in 1876, responsibility for the *jihad* came increasingly to be a concern of the individual and local community, rather than the state. The institution of a personal *jihad*, called *juramentado* by the Spanish, was a form of suicide in which a man went to a Christian settlement and ritually began to murder non-Muslims until he in turn was killed. (Kiefer 1973: 108)

On the second level of meaning, the *Sinrili'* is a version of an ancient Austronesian myth in which an original androgynous whole is split into a pair of opposite sex twins. The twins are separated at birth and spend their lives trying to reunite. The ultimate reconstitution of the androgynous whole results in a burst of regenerative power that produces a new cycle of life. This scenario underlies accounts of the origin of the cosmos throughout Indonesia. It also underlies the myths known in Java and Bali as the story of Panji and in South Sulawesi as the I La Galigo, myths in which a prince spends his life roaming the forests and seas trying to reunite with his twin sister and often ending up with an identical substitute instead. In the royal foundation myths of the Makassar, the role of the male twin is played by a sea-faring prince from an existing political center such as Majapahit or Bantaeng, while the role of the female twin is played by a semidivine princess who has descended from the heavens. Their union gives rise to a class of nobles who agree to provide a local population with peace and order in exchange for tribute and obedience (see Gibson, 2005 for a complete analysis of these myths).

In the *Sinrili'*, the pair of twins is separated not so much by geographic space as by social distance, in the Datu Museng is a poor orphan and Maipa Deapati is a high-born princess. Their physical union results in their death, but their death also results in their mystical union. Instead of producing a local class of rulers, they produce a joint tomb of immense power. Just like the tombs of Shaikh Yusuf and Daeng Nisanga, the tombs of Datu Museng and his "sister"/wife Maipa Deapati remain popular pilgrimage sites to this day for newly weds. The tombs of the latter have a slightly lesser status than those of the former. They are the objects of veneration for an entire ethnic group rather than an entire religious community. But they still demonstrate the ability of charismatic power and knowledge to transcend the boundaries of all kinship ties and create a community united by resistance to and the transcendence of a corrupt colonial order.

This level of meaning allows the *Sinrili' Datu Museng* to serve as a mythical charter for noble wedding rituals. In the village of Ara, it is usually recited during the three or four nights that precede a noble wedding, the most elaborate and emotionally charged ritual in Makassar life. The bride and groom are in a precarious spiritual state during this period and close family and friends must maintain a sleepless vigil over them throughout the nights. Similar vigils are held for the mourners in the days following a funeral. The audience includes both the wedding guests and the couple who are about to undergo the most important transformations in their social and emotional lives. At a symbolic level, Makassar wedding rituals represent every wedding as the predestined reunification of an ancestral sibling set through cousin marriage; as an elopement in which an ambitious man overcomes the defenses of a high-ranking woman and her parents; as a formally negotiated alliance between autonomous houses; and as a mystical bonding between individual souls that is accomplished through an emotionally satisfying physical consummation.

On the third level of meaning, Datu Museng's four journeys may be reinterpreted in light of the four stages of the mystical path and of the human life cycle as understood in Sufi thought. At this allegorical level, the relationship between God and Man is symbolized not by a Father willing to sacrifice His son as in the Abrahamic traditions of the Middle East, but by a pair of lovers, a common trope in many mystical traditions. Life is a long struggle to know and to reunite with the Creator Who underlies all creation. Datu Museng's four journeys can thus be reinterpreted as steps on the path to God. The journey from Taliwang to Sumbawa Lompo represents the acquisition of the *shariah*, the external laws of religion, for that is where he learns the Koran and feels the first attraction to Maipa. The journey from Sumbawa Lompo to Mecca represents the acquisition of *tariqa*, the inner meaning of the *shariah*, for that is where he is instructed by the spiritual masters of Islam and acquires the esoteric knowledge necessary to return to Sumbawa Lompo, penetrate the palace, and physically unite with Maipa. The journey from Taliwang to Ujung Pandang represents the realization of *haqiqa*, the intuition of ultimate reality, for it is during this sea crossing that the lovers vow to pursue a deeper union beyond the physical bounds of this life. The journey from Ujung Pandang to heaven represents the passage to *marifa'*, gnosis. The brother/husband is buried on the shore and the sister/wife is buried inland, but after seven days, their tombs miraculously merge. Lover and Beloved, sea and land, male and female merge back into the primordial androgynous unity. Their tombs lie at the center of the city of Ujung Pandang and continue to serve as a source of spiritual blessings for devotees to this day.

Conclusion: Epic Narrative as *Bricolage*

Datu Museng's life can be read both as a rebellion against an illegitimate colonial order, as a transformation of the Austronesian origin myth, and as statement on the spiritual irrelevance of all political orders. The fusion of these levels of meaning in the *Sinrili'* provide an intensification and transformation of the model provided by Shaikh Yusuf. Where Shaikh Yusuf spent the last sixteen years of his life teaching the mystical path under Dutch surveillance, Datu Museng chose death rather than submission to infidel power. The eruption of extra-social forces brings death and destruction to all. The entire social order is shaken as the governor himself loses his life. His story links otherworldly mysticism to political resistance in a radically new way.

The subversive political implications of the Tagalog texts discussed by Ileto are hidden beneath a veneer of Christian piety. Nevertheless, they stress that obedience to God was a higher duty than responsibility to family, they denigrate the value of social status based on wealth and education, and they make the hero of the story a poor, humble carpenter. In times of political upheaval, they provided peasants with a model for a revolutionary renunciation of their obligations to their overlords.

> I am not suggesting that the masses drew a one-to-one correspondence between pasyon images and their oppressed condition, although this may in some instances have been the case. What can be safely concluded is that because of their familiarity with such images, the peasant masses were culturally prepared to enact analogous scenarios in real life in response to economic pressure and the appearance of charismatic leaders. (Ileto 1979: 24)

The subversive implications of the Makassar texts I have discussed are even more obvious. The Islamic heroes are also men of humble birth who use religious knowledge to openly challenge corrupt VOC officials and arrogant local kings.

One difference between the Tagalog and Makassar cases is that the Spanish Inquisition deliberately destroyed much of the pre-Christian myth and literature of the Philippines, while the pre-Islamic myth and literature of Indonesia survived. As a result, it is easier in the Makassar case to see how epics represent a synthesis of ancient Austronesian myth, early colonial history, and monotheistic theology. By tracing the development of these epics from the seventeenth century to the present, we can see the kind of symbolic work Lévi-Strauss called *bricolage* in action (Lévi-Strauss [1962] 1966).

Chapter 5

Popular Mysticism and the Colonial State, 1811–1936

In this chapter, I explain how Arung Palakka, the Bugis noble from Bone who allied himself with the VOC to defeat Sultan Hasan al-Din of Gowa in 1667, was transformed into Andi' Patunru, the hero of a popular Makassar narrative called the *Sinrili' Tallumbatua*, or *Epic of the Three Boats*. This transformation occurred in tandem with the decline of the royal houses of Gowa and Tallo' and the continued prosperity of the royal house of Bone during the eighteenth century. By 1786, Sultan Ahmad al-Salih of Bone (1776–1812) had expanded his influence over all of South Sulawesi and had persuaded many Makassar that he was the rightful heir to the throne of Gowa. This enabled subsequent generations of Makassar to overlook the preceding conflicts between the royal houses of Gowa and Tallo' and to regard Arung Palakka as a Makassar prince.

The Government of the Netherlands East Indies fought three wars (in 1824, 1859, and 1905) to subdue Bone. As the power and prestige of all the kings declined during the nineteenth century, the *shaikhs* of the Sufi orders moved out of the royal courts and into the villages of South Sulawesi. The most popular of these orders was the Sammaniyya. Like the many of the other "neo-Sufi" orders in the eighteenth century, its *shaikhs* combined a rigorous training in the mystical *tariqa* with an equally rigorous study of the *hadith*. Immersion in the *hadith* tended to replace devotion to one's Sufi master with devotion to the Prophet Muhammad. This was expressed through the collective recitation of devotional texts such as the *Maulid al-Nabi* of Jaffar al-Barzanji, or *Barasanji*. The *Barasanji* presents a distinctly populist image of Muhammad as a poor orphan whose charismatic power was acknowledged by all the great kings of his age.

I argue that it was this portrayal of the Prophet as a charismatic orphan that provided the model for the *Sinrili' Tallumbatua's* portrayal of Arung Palakka' as a pious Makassar prince who was forced to flee his homeland by an arrogant king. By the time the *Sinrili' Tallumbatua* was recorded in 1936, the colonial state had taken over the political functions of the royal courts of South Sulawesi, while popular Islamic institutions had taken over their religious functions. It was only after this separation between state and mosque had been accepted by most Makassar that it was possible for them to view Arung Palakka's alliance with the VOC in the seventeenth century as part of a pragmatic strategy rather than as an act of apostasy.

Origins of the Early Colonial State, 1811–1860

Keeping the kingdoms of Gowa and Bone separate and at odds with one another was the cornerstone of VOC policy in South Sulawesi (von Stubenvoll 1817 II: 12, 28). As we saw in chapter 4, the strongest political leaders in South Sulwesi between 1739 and 1760 were Sultan Shafi al-Din of Tallo' and his wife and first cousin Amira Arung Palakka. Amira was the daughter of Sultan Ismail of Bone by his Gowanese wife, Karaeng Pabineang, who was the sister of Sultan Siraj al-Din of Gowa. Amira was thus eligible for the thrones of both Bone and Gowa in her own right, and successive Dutch governors were constantly anxious that she would manage to unite the two kingdoms under a single ruler.

Like Shafi al-Din, Amira Arung Palakka was deeply hostile to the Dutch presence in Sulawesi (von Stubenvoll [1759] 1817 II: 20). Shafi al-Din and Amira never claimed the throne of Gowa themselves, but ruled shrewdly through their children and grandchildren. In 1735, they placed their twelve-year-old son, Abd al-Khair (r. 1735–1742), on the throne of Gowa. At the height of Bontolangkasa's uprising in 1738, they were able to place their daughter Siti Nafisah on the throne of Bone. When Abd al-Khair died in 1742, they placed their nine-year-old son, Abd al-Qudus (r. 1742–1753), on the throne of Gowa. When Abd al-Qudus died in 1753, they placed their four-year-old grandson, Amas Madina Batara Gowa Fakhr al-Din (r. 1753–1767) on the throne of Gowa.

When Shafi al-Din died in 1760, Batara Gowa was only eleven years old and Amira Arung Palakka exercised effective power. When Batara Gowa reached the age of seventeen in 1766, he was able to govern without a regent. Almost immediately, however, he abandoned his kingdom and sought refuge with his mother in the kingdom of

Bima. He was captured there by the VOC in 1767 and exiled to Sri Lanka, where he died in 1795. Batara Gowa was replaced by his brother, Arung Mampu Madu al-Din, but he too abandoned the throne in 1769. He was under the influence of his grandmother, Amira Arung Palakka, and she may have played a key role in persuading him to abdicate. At this point the VOC saw to it that Arung Mampu was replaced as sultan of Gowa by Zain al-Din (r. 1769–1777), a younger brother of Shafi al-Din with no ties to Bone.

In 1776, after thirty-seven years during which only young figureheads had occupied the throne of Gowa, a charismatic imposter called Sangkilang suddenly appeared on the scene. He claimed to be the exiled king, Batara Gowa, and quickly gathered an army of followers. Among them were the real Batara Gowa's grandmother Amira Arung Palakka; his brother, Arung Mampu; and his aunt, Siti Salaha II the future queen of Tallo' (r. 1780–1824). All of them recognized Sangkilang's authenticity. Sangkilang's forces captured the royal palace of Gowa in June 1777, and he was acclaimed as king. The VOC did not manage to expel him from Gowa until June 1778. He escaped to the mountains with the Sudang, the sacred sword that had been left behind by Gowa's founding ancestor, Lakipadada. He was accompanied by the now aged Amira Arung Palakka, who remained at his side until her death in 1779.

The Dutch had meanwhile proclaimed Zain al-Din's son, Abd al-Hadi (r. 1778–1810), as the sultan of Gowa. But without the Sudang sword, and with Sangkilang still at large, Abd al-Hadi's legitimacy was tenuous. Abd al-Hadi was only formally installed as sultan of Gowa after Sangkilang died in 1785. The latter left the regalia to the former sultan of Gowa, Aru Mampu, who turned them over to Sultan Ahmad al-Salih of Bone (Buddingh 1843: 450–451). Ahmad al-Salih's claim to be a Makassar prince with a right to possess the "royal ornaments" of Gowa was based on his descent from Sultan Ismail, who had ruled Gowa from 1709 to 1712, and Bone from 1720 to 1721 (van Hoevell 1854: 167; Tideman 1908: 359; but see Bakkers 1866: 163, 165). By the time Sultan Ahmad al-Salih came to the throne of Bone in 1775, the two royal lines were so intertwined that most of the Makassar living in the mountains of Gowa had no difficulty in accepting Sultan Ahmad al-Salih of Bone as a legitimate heir to the throne and regalia of Gowa (see figure 5.1). After the death of Sangkilang in 1785, most mountain Makassar recognized Ahmad al-Salih as their ruler, while most coastal Makassar recognized Abd al-Hadi.

The VOC entered into its final decline in 1780 when the Netherlands became embroiled in the Fourth Anglo-Dutch War as a result of Dutch

114

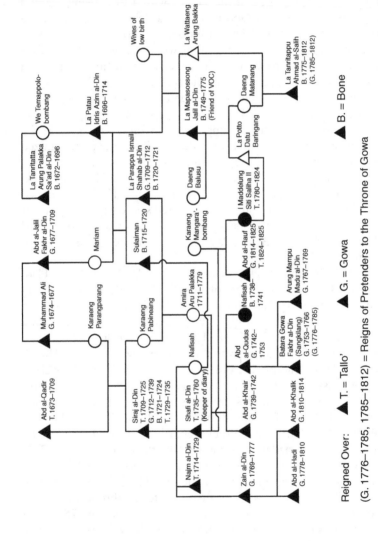

Figure 5.1 The Merging of Gowa and Bone, 1672–1812

Reigned Over:　▲ T. = Tallo'　▲ G. = Gowa　▲ B. = Bone

(G. 1776–1785, 1785–1812) = Reigns of Pretenders to the Throne of Gowa

support for the American war of independence. For the next four years, VOC authorities in the East Indies were starved of resources and were unable to eliminate rebels such as Sangkilang. When Sultan Ahmad al-Salih of Bone refused to surrender the regalia of Gowa in 1785, preparations were made to dispatch a fleet from the Netherlands to punish him. The expedition was cancelled in 1787 by supporters of Prince William of Orange who feared antagonizing their British allies.

When the French Revolution broke out in 1789, news of turmoil in Europe soon spread to the East Indies. In 1794, Sultan Ahmad al-Salih invaded the Northern Districts of the VOC and established a new court at Rompegading. In 1795, Dutch supporters of the French Revolution seized power in Amsterdam and declared the Batavian Republic. In 1797, Sultan Ahmad al-Salih concluded a secret agreement with the British and attacked the Dutch bases in Bantaeng and Bulukumba. When Napoleon declared his brother, Louis, King of Holland in 1806, VOC territories throughout Asia became subject to British attack as the possessions of a ruler allied to the enemy.

When the British decided to occupy the Dutch East Indies in 1811, there was little resistance. Van Wikkerman surrendered Ujung Pandang without a fight at the beginning of 1812. Sultan Ahmad al-Salih thought that the time was at hand for the restoration of Muslim authority over all of South Sulawesi. He soon discovered, however, that the British had other plans. Like the Dutch, they pressed him to surrender the Sudang sword so that they could use it to install a compliant ruler of their choice in Gowa. But Sultan Ahmad al-Salih refused to abandon his own desire to bring the throne of Gowa under the control of Bone.

Ahmad al-Salih died in 1812, but his son, Sultan Muhammad Ismail (r. 1812–1823), continued his father's policies. In 1814, the British drove the forces of Bone out of Gowa and installed Abd al-Rauf (r. 1814–1825) as the sultan of Gowa (de Klerck 1938 II: 50). At long last, the British managed to gain possession of the Sudang sword. Before the British could restore the Sudang to the new sultan of Gowa, the Napoleonic wars suddenly came to an end in Europe. The British were preoccupied with the creation of a strong state to the north of France. They incorporated Belgium into the Netherlands and returned most of the overseas territories they had taken from the VOC. According to the Constitution of 1814, these territories became colonies under the direct control of the king of the Netherlands (de Klerck 1938 II: 71–73).

In the east, the British had no desire to expand the territory under their direct administration. They did want their merchants to have free access to ports everywhere in Asia, however. South Sulawesi was transferred to the new Dutch colonial government in April 1816. One of its first acts was to return the Sudang sword to Abd al-Rauf and to formally install him as the sultan of Gowa (r. 1814–1825). The mountain Makassar who had declared their allegiance first to Sangkilang and then to Sultan Ahmad al-Salih of Bone were finally reconciled to rule by the sultan of Gowa. The colonial government spent the next five years trying to persuade all the other local rulers to renew the treaties they had signed with the VOC.

The mercantilist model followed by the VOC rested on the monopolization of trade, the use of bound and slave labor, and the forcible extraction of tribute. The VOC also allowed Dutch officials to form stable relationships with local women, generating a large class of Indo-Europeans who served as linguistic and cultural brokers. According to the liberal model of colonialism the British introduced during their occupation of the Dutch East Indies, native society was to be transformed through the introduction of free markets in land and labor, and a strict social boundary was to be maintained between ruler and ruled. The British found repugnant the eighteenth-century Dutch practice of keeping native concubines who continued to adhere to native customs. They set a new social standard of drawing native and Indo-European wives into a European social life of polite conversation, dancing, and dinners. As the century wore on, a preference grew for bringing wives out from the motherland and for drawing sharper racial distinctions in the colonies (Fox 1985; van der Veer 2001; J. Taylor 1983).

The British had instituted a policy of free trade during the occupation of the Dutch East Indies, but it had lasted for only two years after they left. After 1818, foreign ships were only allowed to trade at Batavia, and even then they were taxed at a higher rate than Dutch ships. The only exception to this rule was that two Chinese junks were allowed to trade for tripang in Makassar each year (Heersink 1995: 78–79). When the British founded Singapore as a free port in 1819, it quickly became the principal source of European goods throughout the archipelago and Batavia became a backwater. The Dutch finally opened Makassar to British shipping in 1848. Makassar's trade with Europeans did grow rapidly between 1847 and 1873, but only as a transit harbor between eastern Indonesia and Singapore (Heersink 1995: 98).

When the Dutch continued to require the subjects of allied and vassal states such as Bone and Wajo' to acquire passes and to pay taxes in order to trade in Makassar, Sultan Ahmad Saleh of Bone imposed a boycott of the port. The only local people who did take advantage of

Makassar were from areas such as Bantaeng and Selayar that were under direct colonial rule. It was relatively easy for them to obtain the requisite passes and they were taxed at a lower rate than the subjects of independent kingdoms (Heersink 1995: 78–79). Most Bugis merchants from Bone and Wajo' found it more profitable to trade in the truly free port of Singapore and in the independent sultanate of Sulu in the Philippines.

As tensions rose between Sultan Ahmad Saleh and the Dutch, the Dutch looked for a highly placed noble they could use to undermine his authority. They found him in Ahmad Singkarru Rukka, an estranged brother-in-law of Sultan Ahmad Saleh of Bone (Bakkers 1866: 167–189). When the queen of Bone, Basse Arung Kajuara ordered Bone ships to fly the Dutch flag upside down in 1858, the Dutch put together a punitive expedition. Ahmad Singkarru Rukka was placed in command of a number of Bone's troops, but failed to put up a credible resistance against the Dutch attack. He was suspected of treason and fled to Barru, where he made a deal with the Dutch. He promised to raise a Bugis army in Sinjai, where he had formerly been the ruler of Bulo Bulo. In return, the Dutch promised to give him Kajang and Sinjai in feudal tenure once they were taken from Bone (de Klerck 1938 II: 313). A second Dutch expedition set out in November 1859 and occupied Bone. Basse Arung Kajuara fled inland to Soppeng and the Dutch installed Ahmad Singkarru Rukka as Sultan Ahmad Idris of Bone. He signed a treaty on February 13, 1860, renouncing all claims to Sinjai, Kajang, and Bulukumba to the colonial government.

Popular Mysticism and the Late Colonial State, 1860–1935

As we have seen, the royal house of Bone cultivated its traditional authority during the eighteenth century by assiduously intermarrying with the other royal houses in South Sulawesi. It cultivated its charismatic authority at the same time by intermarrying with the descendents of Shaikh Yusuf. In 1706, Sultan Ismail (r. Gowa 1709–1712, r. Bone 1720–1724) married Labiba (Gumitri), a daughter of Shaikh Yusuf by his wife Kare Kontu (Ligtvoet 1880: 178). A generation later, Sultan Jalil al-Din (r. 1749–1775) married Habiba, the daughter of Shaikh Yusuf's son Muhammad Jalal by his wife Aminah, who was herself the daughter of Sultan Ageng of Banten (r. 1651–1683) (Cense 1950: 54; Buyers 2000–2005).

Sultan Ahmad al-Salih (r. 1775–1812) cultivated his own charismatic authority by personally translating several of Shaikh Yusuf's writings into Bugis during the 1780s and by maintaining a circle of scholars and mystics at his court that translated many other Islamic texts from Arabic and Malay into Bugis (Cense 1950: 54–56). He appointed a Javanese from Bogor called Yusuf as the *kali* of Bone. Yusuf Bogor initiated the sultan into the reformed branch of Shaikh Yusuf's Khalwatiyya founded by Muhammad al-Samman. Ahmad al-Salih restricted the study of mysticism to members of the royal court, and there is no record of Yusuf Bogor initiating anyone but the sultan himself into the Sammaniyya (van Bruinessen 1991: 260).

Despite his opposition to the popular study of mysticism, Ahmad al-Salih set into motion a number of processes that facilitated the popularization of mysticism during the nineteenth century. His own initiation into the Sammaniyya legitimated a new kind of mysticism among the nobles of Bone. The translation of Islamic texts into Bugis obviously made them more accessible to the general population.

As the wealth and power of kings declined in the face of the expanding power of the colonial state, the courts were replaced as the center of religious life by rural schools called *pesantren*. Martin van Bruinessen has described this process in Banten during the nineteenth century.

> In the heyday of the sultanate, Islamic education took place at the centre, under the sponsorship of the kraton, and with members of the royal family among its chief beneficiaries. . . . This changed with the decline and ultimate demise of the sultanate. Independent teachers emerged in the periphery. Snouck Hurgronje made the important observation that *zakat* began flowing to these independent ulama rather than the Dutch-appointed *pangulus*, but perhaps the emergence of the independent ulama as a group reflects some earlier shift in economic resources enabling certain families to send one or more relatives to Mecca for studies. The number of *pesantrens* rapidly increased in the late 19th century. At the same time the *tarekat* Qadiriyya wa Naqshbandiyya gained a mass following in the villages. (van Bruinessen 1995: 191–192)

A similar process occurred in South Sulawesi, where the royal court of Bone also declined as a center of Islamic learning after the death of Sultan Ahmad al-Salih. During the nineteenth century, the Sammaniyya played the same role in South Sulawesi that the Qadiriyya wa Naqshbandiyya did in Java.

Figure 5.2 illustrates the way influences from the cosmopolitan networks of Shafii *ulama* and Khalwati *shaikhs* were continually intro-duced into South Sulawesi from the time of Haji Ahmad al-Bugisi

119

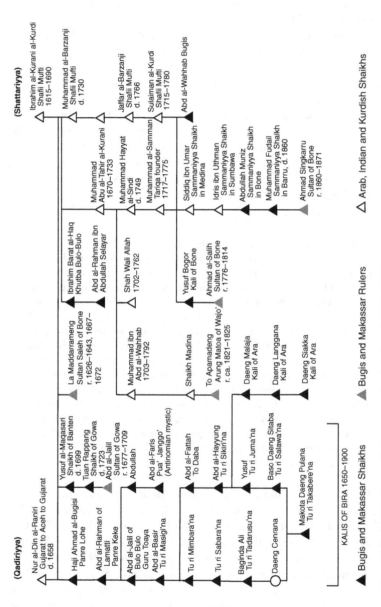

Figure 5.2 Middle Eastern Influences on South Sulawesi, 1650–1850

in the 1650s until the time of Abdullah Muniz in the 1850s. It was through these networks that the recitation of al-Barzanji's *Maulid* and of the Sammaniyya's loud public *dhikr* was introduced into village ritual all over Indonesia.

The most common development in the eighteenth and nineteenth centuries was the combination of intensive study of the *hadith* and the pursuit of the *tariqa*. This resulted in a close mystical identification with the Prophet as opposed to Sufi *shaikhs* (Voll 1987: 87). The renewed emphasis on the example of the prophet as interpreted by popular *ulama* and *shaikhs* was also due in part to the progressive delegitimation of Muslim rulers throughout the Islamic world by the steady expansion of European power during the eighteenth century. The Arabian provinces of the Ottoman Empire were among the first areas to experience this expansion. Muhammad al-Samman was himself initiated into the Khalwatiyya by Mustafa al-Bakri (d. 1749), a Syrian *shaikh* from Damascus.

> The revival of the Khalwatiya in Egypt, inspired by al-Bakri and sustained by his friend and pupil al-Hifnawi (d.1768), coincided with the stirrings of reform and change not only in Egypt, but elsewhere in the Islamic world. Some of these changes were accelerated by European encroachments and pressures, others by the visible decay of the Ottoman Empire. . . . As the rate of troubles and exactions accelerated, the people looked increasingly to the lower ranks of the ulama and to the *shaykhs* of the brotherhoods for redress against oppression and plundering. . . . The *shaykhs* were able to take on this role because they were financially independent. Many of their *zawiyas* were supported by *waqfs* which were hard to plunder, and the *shaykhs* continually received gifts and presents from the population, and sometimes from foreign rulers. Thus their standing was unassailable and they were immune to pressures. They were closer to the people, and as part of their "pastoral" function, listened to their complaints and advised them whenever they could. (Martin 1972: 298–299)

During his time in the Hejaz, Yusuf Bogor came into contact with a number of other reformist influences. He was initiated into the Shattariyya order by Muhammad Tahir al-Kurani (1670–1733), a *shaikh* who also taught the great Indian reformers Shah Wali Allah and Muhammad Hayya al-Sindi (van Bruinessen 1991: 259–260). Muhammad Tahir was the son of Ibrahim al-Kurani (S21, 1616–1690). As we saw in chapter 3, Ibrahim al-Kurani was the Shafii *mufti* of Medina from 1661 until 1690 and taught an earlier generation of Southeast Asian students, including Abd al-Rauf of Aceh (ca. 1620–ca. 1693), Shaikh Yusuf of Makassar (1626–1699), and Ibrahim Barat of Bulo Bulo (van Bruinessen 1987: 47).

Ibrahim al-Kurani was only the first of a series of Kurds who served as the *mufti* of the Shafii school of law in Medina between 1660 and 1780. Since they adhered to the Shafii school of law, advanced Indonesian students often studied under these Kurdish *mufti*. This accounts for the otherwise inexplicably strong Kurdish influence on Indonesian Islam (van Bruinessen 1987). Ibrahim al-Kurani was succeeded as Shafii *mufti* by another Kurd, Muhammad ibn Abd al-Rasul al-Barzanji (d. 1730). Muhammad al-Barzanji is best known for his vigorous defense of Ibn al-Arabi's doctrine of the Unity of Being against Sirhindi's doctrine of the "Unity of Witness," *wahdat al-shuhud*. Muhammad wrote two books attacking Sirhindi, claiming that the latter had gone so far as to blasphemously imply that he was a new prophet. Some of the Indians living in Mecca defended Sirhindi against this criticism in 1683 (Rizvi 1983: 339–441; van Bruinessen 1987: 48).

Muhammad al-Barzanji was succeeded as Shafi *mufti* by his grandson, Jaffar al-Barzanji (1690–1766). Jaffar al-Barzanji was the author of the *Maulid al-Nabi*, a poetic life of the Prophet Muhammad that became popular from West Africa to South Sulawesi. (Knappert 1971: 48–60; for an Indonesian translation, see Abu Aufa ash-Shiddiquie 1986). In South Sulawesi, it is usually referred to as the *Barasanji* and became central to ritual life during the nineteenth century. What is most striking about the *Barasanji* is its emphasis on the humility of the greatest of the prophets. Despite an illustrious patrilineage that includes many of the prophets, the text stresses that Muhammad was born without a father and was raised by four different women. Aside from his birth mother, they included a slave, a beggar, and a servant. He later married a woman who was older, wealthier, and more powerful than himself. He endured the scorn and contempt of even the lowest ranking of the Meccans without anger or violence. And yet he had no fear of the most powerful of kings, whose mighty cities trembled at his birth and whose learned men feared the portents that accompanied it. The signs of his prophecy were clear to all the priests versed in esoteric knowledge. He was a humble hero who used his transcendental power to defend orphans, widows, and the poor against the arrogant and corrupt rulers of his time.

The moral lessons contained in the text are clear: prophetic knowledge and the charismatic power that goes with it is separate from, superior to and in some respects an inversion of ordinary social rank and political power. The moral character of the poor and oppressed is often superior to that of the rich and powerful. Women are often capable of greater compassion and benevolence than men. Normal processes of aging and corruption are inverted for prophets and *shaikhs*: their sweat and their corpses are fragrant and life giving. In sum, ordinary social hierarchies based on gender, class, and power are shown to be irrelevant to the spiritual hierarchy, which is based on moral rectitude and charismatic blessing alone. These lessons help explain why the *Barasanji* became so popular in the villages of South Sulawesi and in other parts of the Islamic world, especially after European expansion had begun to undermine the authority of local rulers.

Jaffar al-Barzanji was succeeded as *mufti* by yet another Kurd, Sulaiman al-Kurdi (b. 1715, *mufti* 1766–1780). Many of the Indonesians who were initiated into the Sammaniyya by Muhammad al-Samman studied Shafii law under Sulaiman al-Kurdi. Sulaiman sent four influential Indonesian students home in 1772 to spread his teachings: Abd al-Rahman of Batavia; Abd al-Samad of Palembang; Abd al-Wahhab Bugis; and Muhammad Arsyad of Banjarmasin (van Bruinessen 1987: 50). Muhammad Arsyad and Abd al-Wahhab Bugis had also studied mysticism under Muhammad al-Samman. Al-Samman's other Indonesian students included Masri of Batavia; 'Abd al-Samad of Palembang, who introduced the Sammaniyya to Sumatra; and Yusuf Bogor, the *kali* who introduced the Sammaniyya to Bone. We can see from these interpenetrating networks that the study of the *hadith* that underlay the *shariah* law continued to go hand-in-hand with the study of Sufi *tariqa* until the end of the eighteenth century.

One response to the renewed emphasis on the study of the *hadith* in the cosmopolitan networks was an increased skepticism of all mystical doctrines and practices that had arisen since they were compiled in the ninth century. Thus the same man, Muhammad Hayya al-Sindi, taught both Muhammad al-Samman, who founded a reformed Sufi order, and Muhammad ibn Abd al-Wahhab (1703–1792), the epony-mous founder of Wahhabism. The Wahhabis turned decisively against all forms of Sufism, especially the veneration of prophets and saints (Voll 1988). The Wahhabi conquest of Mecca in 1803 inspired three *hajjis* who returned to Sumatra and became leaders in the Padri reform movement (Ricklefs 1981: 133–134). In the 1820s, a follower of Ibn abd al-Wahhab known as Shaikh Madina became the religious adviser of the Arung Matoa of Wajo', La Mamang Toapamadeng (r. ca. 1821–1825). Just as the Wahhabis had destroyed the tomb of the Prophet in 1803 as a center of idolatry, so the followers of Shaikh Madina in Wajo' destroyed the trees that housed the nature spirits and the shrines that housed the ancestor spirits (Mattulada 1976: 71–72).

A third response to the growing stress on legal study was an antinomian repudiation of the *shariah* altogether by local mystics. This is what seems to have happened in Bira at the end of the eighteenth century when a descendent of Haji Ahmad called Abd al-Haris forsook the mosque and began meditating on top of a local mountain peak that overlooks the Strait of Selayar. He is known to his devotees as Pua' Janggo', The Bearded Master, and is held to have acquired enormous supernatural powers. His teachings were condemned at the time by his first cousin, Abd al-Basir, who is known by his descendents as Tu ri Masigi'na, "The One of the Mosque." The division between the "party of the mosque" and the "party of Pua' Janggo' " persists in Bira to this day. The followers of Pua' Janggo' pursue their mystical exercises in secret because of their heterodox character. But many ordinary people openly visit the *karama'* site where Pua' Janggo' used to meditate. They seek his intercession in cases of illness and other misfortunes, and make vows, *nazar*, that they promise to fulfill when their requests are granted.

The Sammaniyya in South Sulawesi

As a popular Sufi order, the Sammaniyya in South Sulawesi traced its origins not to the branch established by Yusuf Bogor in the royal court of Bone, but to a separate branch of the order that derived from Muhammad al-Samman by way of Siddiq ibn 'Umar Khan of Medina. Siddiq initiated a South Sumatran called Idris ibn 'Uthman, who settled in Sumbawa. Idris initiated a number of Bugis disciples there, including 'Abdallah al-Munir, the son of a Bugis noble from Bone. 'Abdallah al-Munir's son, Muhammad Fudail (ca. 1790–1860) settled in Barru and made it the main center of the Sammaniyya in South Sulawesi during the 1850s (van Bruinessen 1991: 260–261). The future sultan of Bone, Ahmad Idris, was initiated into the Sammaniyya by Muhammad Fudail while he was living in Barru during the 1850s.

Muhammad Fudail died in 1860, just as Ahmad Idris was being installed as sultan of Bone. Leadership of the Sammaniyya then passed to Haji Palopo, (d. 1910), a member of Bone's lower nobility. Under his leadership, the Sammaniyya evolved from an order closely tied to the royal court to one with a widely dispersed following centered on its peripatetic spiritual head. Haji Palopo eventually settled in Marusu', where La Madarammeng had founded a Sufi lodge in 1643 and where Shaikh Yusuf's sons had founded a Khalwatiyya lodge in 1683. In later years, La Madarammeng's center was absorbed into a branch of the reformed Khalwati order found by Muhammad al-Samman in the 1760s. By the time of his death in 1910, Haji Palopo had several hundred noble followers all over the province (van Bruinessen 1991: 261). His Sammaniyya lodge in Marusu' came to be identified with the Bugis nobility, much as the Khalwatiyya lodge founded in Marusu' by Shaikh Yusuf's sons had come to be identified with the Makassar nobility during the eighteenth century.

Haji Palopo was succeeded by his son, Haji 'Abdallah (d. 1964). The order grew even faster under Haji Abdallah, who opened it to commoners. Haji Abdullah even admitted women to the order, and allowed them to take part in the communal recitation of *dhikr*. The latter practice drew lurid criticisms of the order beginning at the time of Haji Abdallah's first visit to Bone in the 1910s. It was said that the *dhikr* was practiced in the company of young virgins, who became pregnant as a result.

> The *dhikr* meetings are closed to outsiders, and the participation of women must have titillated people's fantasies. The loudness of the *dhikr* and the violent bodily movements accompanying it, which must have made the fragile wooden and bamboo *musalla* vibrate, may have given further food for suspicion. (van Bruinessen 1991: 263)

By 1918 the growing numbers of Sammaniyya initiates had become evident to Dutch officials such as Eerdmans. By the 1920s, a growing number of critics were accusing Haji Abdallah of pantheism and other heresies. The *controleur* of Marusu' felt compelled to bring him in for an interview in 1924 during which Haji Abdallah defended himself against the accusations of heterodoxy and illicit sexual activities. He said that while he permitted women to participate in *dhikr*, they had to do so quietly and modestly. In the 1970s, the Department of Religious affairs estimated that the Sammaniyya had 117,435 followers in South Sulawesi. Van Bruinessen comments that if the Department's figure of 70,000 followers in Marusu' alone was correct, they made up two-thirds of that district's adult inhabitants (van Bruinessen 1991: 252). Abu Hamid estimated that the Sammaniyya had 259,982 followers in the province as a whole in 1976 (Abu Hamid 1983: 361). If this estimate was correct, they then made up over 10 percent of all Bugis adults in South Sulawesi. By contrast, the orders that were most popular elsewhere in Indonesia, the Naqshbandiyya and the Qadiriyya, had only 10,000 and 5,000 members, respectively (Pelras 1985: 126–127).

Popular Mysticism among the Konjo Makassar

Bira and its neighbors speak a dialect of Makassar known as "coastal Konjo" that is closely related to the "mountain Konjo" dialect spoken in the highlands of Gowa. The coastal Konjo were vassals of Gowa from 1560 until 1667, when they were given to the VOC by the Treaty of Bungaya. According to local genealogies, descendents of a Gowanese noble called Karaeng Mamampang ruled Ara and Bira from about 1450 until about 1670. But this area has also been exposed to a good deal of influence from Bone that tended to grow whenever Gowa or the VOC was weak. During the reigns of Arung Palakka (r. 1672–1696) and his successor Sultan Idris (r. 1696–1714), a noble from Bone called Puang Rangki came to power in Ara. When the VOC reasserted control of the area in 1726, a descendent of Karaeng Mamampang regained power. But when Sultan Ahmad al-Salih of Bone expanded his sphere of influence in the 1780s, a great grandson of Puang Rangki called Salung Daeng Masalo came to power. As we saw in chapter 3, on the religious side, the *kalis* of Bira throughout this period traced their descent back to Haji Ahmad, the Bugis student of Nur al-Din al-Raniri.

The first Bone war of 1824 ended with Bone still in possession of Kajang, but with the colonial government in firm control of Ara, Bira, Tanaberu, and Lemo Lemo. Salung Daeng Masalo was turned out of office by Ruru Daeng Situru, a son of Karaeng Amar Daeng Matoana of Bira. During the same generation, Daeng Malaja became the *kali* of Ara. He was the son of a *kali* of Bira, Abd al-Hayyung, who traced his descent from Haji Ahmad the Bugis. Both Ruru Daeng Situru and Daeng Malaja married sisters of Mandu, the last *gallarrang* of Ara who traced his descent from the royal house of Gowa.

The offices of *gallarrang* and *kali* tended to be passed down from father to son or from elder to younger brother (see figure 5.3). In the absence of a qualified kinsman, the office of *kali* often passed from a man to his son-in-law. During the nineteenth century almost all the *gallarrang* of Ara traced their descent from Ruru Daeng Situru and almost all of the *kali* traced their descent from Daeng Malaja. The two lines intermarried over the generations, creating a cohesive class of rulers and religious officials. The ruler of Ara in the 1880s was known as Baso Sikiri, "He Who Recites *Dhikr*," testifying the fusion of political and religious roles at the time.

The situation in Bira was somewhat different, since descendents of Karaeng Mamampang never lost control of the village. According to local genealogies, Bira was ruled by Karaeng Miri Daeng Selatan from about 1725 to 1780, and by his great grandson Karaeng Amar Daeng Matoana from about 1780 until about 1830. But as we have seen, the *kalis* of Bira traced themselves back to a series of Bugis mystics who settled in the village between 1650 and 1750. It is quite likely that

126

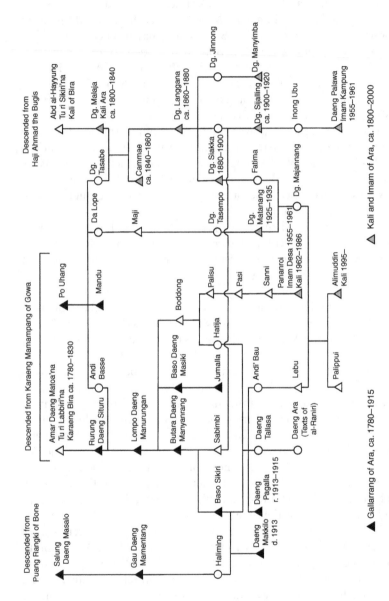

Figure 5.3 *Gallarrang* and *Kali* in Ara, 1780–2000

they stayed in touch with the Islamic scholars and mystics of the Bone court during the reign of Sultan Ahmad al-Salih. And indeed, the *kali* of Bira at the time, Abd al-Hayyung, received the posthumous epithet of *Tu ri Sikiri'na*, "He who Recited *Dhikr*."

Amar Daeng Matoana's son, Baso Daeng Raja, ruled Bira from about 1835 until 1884. Baso Daeng Raja's son, Andi' Mulia Daeng Raja ruled from 1900 to 1914 and from 1921 to 1942. Andi' Mulia was a devout traditionalist who performed the *hajj* in 1914 and who was receptive to the Sammaniyya practices that were spreading down the social hierarchy in the early twentieth century. The British author, G.E.P. Collins, observed an elaborate ceremony in honor of the Prophet's birth, or Maulid, sponsored by Andi' Mulia in the 1930s. The energetic movements and loud *dhikr* he described are characteristic of the Sammaniyya (Snouck Hurgronje 1906: 218–221; Drewes 1992: 78–81).

Andi' Mulia arrived with his retinue at the old mosque after dark, carrying the double spear and trident that were part of the regalia. The women and children sat behind a wooden barrier inside the mosque. The women were dressed in silk sarongs and *baju bodo*, a transparent blouse customary all over South Sulawesi. They wore gold chains around their necks, as did the children. The western end of the mosque, where the pulpit stands, was hung with red cloths. Andi' Mulia sat to the right of pulpit, along with the *kali* and thirteen assistants dressed in white coats and round caps woven of colored grass or cane. A yellow scarf with red tassels was wrapped around each cap. The former regent, Baso Daeng Makanyang (r. 1920–1931), sat near Andi' Mulia. The other chiefs sat to the left. Red, yellow, black, and white checkered cloths were hung above their heads. Over the rest of the assembly, coconuts, bananas, watermelons, oranges, maize, sugar cane, and paper decorations were suspended. Some of these were attached to wide rattan frames called *gintung*. There were also little paper houses full of cakes, bananas, maize, and other foods, and angular biscuits at the end of long strings.

An elaborate offering was laid out in front of the chiefs. Some were placed in deep baskets, others on large brass trays with high pedestals (*kapparra bangkeng*). Long pieces of dried black fish had been woven into globes. The regent's globe was two feet in diameter. Spongy omelets had been placed on top of the globes.

The *kali* and his assistants began to chant *dhikr*, alternating with a chorus of women. This went on for two or three hours. Whenever the Prophet's name was mentioned, the men shouted out "*Allahu ma sallih!*" "God loves him!" At the end of each group's singing, they shouted out "*Sallalahu allahi wah salam*," "God loves him and grants him good fortune." The songs were read from "old Arabic books," "not from the Koran." This is probably a reference to the *Barasanji*. While this was going on, Collins talked about geopolitics with Andi' Mulia, Daeng Makanyang, the former *kali*, and the current *kali*, Haji Andi Rukha Muhammad Amin. Eventually, coffee was brought, and the singing paused for a while.

> At about midnight the Kalif's men quickened their singing. Again and again from their corner came a riotous rollicking chorus:
> 'Allah! Ah! Hu! Allahu! Ah! Hu! Ah! Allahu! . . . .
> And with every syllable their bodies jerked up and down, this way and that. Now and then a voice rose to a shriek and its owner looked up at the roof, flinging his arms above his head in ecstasy. Faster and faster they sang, and ever louder. And even when their turn was ended and the women and children began to sing the jerking spasms of the rhythm would not leave their bodies. . .
> Then I turned to the women and children, in front of the Kalif's men. All sat still and quiet. Not one of them showed any sign of the frenzied emotion that filled the men before them. (Collins 1936: 164)

They chanted on for "long hours," until there was another short rest. Then those who were still awake lunged for the hanging foods and consumed them. The baskets and trays on the floor were later taken home by their owners. Then the singing began again and continued to dawn, marred only by the collapse of a wall that gave way under the pressure of five singers (Collins 1936: 157–169).

The recitation of the *Barasanji* was an integral part of both the aforementioned *dhikr* sessions and of most life-cycle rituals. Pelras has suggested that among the Bugis the recitation of the *Barasanji* was substituted for the recitation of appropriate episodes from the I La Galigo during noble life-cycle rituals (Pelras 1996: 203). This makes a good deal of sense: while much of the underlying ritual symbolism in these life-cycle rituals remained largely intact, the Prophet Muhammad replaced the gods of pagan mythology as the explicit model for noble conduct.

Collins provides a description of an elaborate performance of the *Barasanji* held in the hamlet of Kasusu on the first anniversary of a man's death. The house of the dead man's family was decorated with blue, red, and yellow cloths. Bananas, sugar cane, cakes, bits of leaves, coconuts, and cylindrical biscuits were suspended from ropes. Banana stems were set in the floor, and hundreds of thin, two-foot strips of bamboo were fixed in them. *Uhu-uhu* and *dumpi* cakes, egg shells, and cut paper decorations were placed on the skewers. The *kali* and his retinue of mosque officials dressed in white coats and woven caps sat at the back. They formed part of a circle of twenty men. Nearby was another circle of twenty men from Ara. In front of the *kali* was a bronze bowl with twenty candles on its rim. He crumbled some incense into a burner and began to pray. Then a long string of heavy beads was passed round until all forty men were holding it. The *kali* shouted "*La illaha ill'allah*," "There is no God but God." At each accented syllable the beads were pulled a little from left to right. The men took up the shout, and the pace speeded up. After half an hour of vigorous shouting and arm flapping, the *kali* gave a cry and threw up his hands, at which all released the string.

Then the *kali* and some others began to read from "Arabic books," again most probably the *Barasanji*. While the *kali* and an old man carried on their recital, the "mosque officials" began a long mournful chant. Then, at a certain moment, all leapt to their feet and grabbed the sweets down from the ropes. They then sat and the two circles of men began to sing in turns, each trying to out-do the other. When this was finished, coffee was served and then rice and curried buffalo. When they finished eating, the eggshells and strips of cut paper were handed round. The paper was licked and stuck to their foreheads. New candles were stuck to the bowl, and more incense was burned. The *kali* recited more prayers, and the choruses started up again. The men flapped their arms and jerked their heads to the rhythm until all stood up. The rhythm became faster and faster and the men began to leap in the air with wild cries. "At last the Kali held up his hands, and the noise ceased with final mad leaps and yells" (Collins 1937: 64–69).

The Epic of the Three Boats as Populist Allegory

In 1670, the royal scribe to Sultan Hasan al-Din of Gowa had represented Arung Palakka, who had been held responsible for the fall of an Islamic kingdom to an infidel army, as follows: "Hear now, companions and friends, of the ill-favored Bugis of uncertain sense [Arung Palakka], whose friendship with Dutchmen, will one day choke him to death" (My translation of stanza 149 of the *Sja'ir Perang Mengkasar*, Skinner 1963: 112).

When many Makassar shifted their allegiance from the royal house of Gowa to that of Bone in 1785, the role played by Arung Palakka in the fall of Gowa began to undergo a revision. By the early twentieth century, Arung Palakka had been fully rehabilitated and was portrayed as a Makassar folk hero in the *Sinrili' Tallumbatua*, or *Epic of the Three Boats*. The name used for Arung Palakka in the *Sinrili'*, Andi' Patunru, means "The Conquering Prince." In 1936, A.A. Cense commissioned a written version of this epic as it was then recited in Jongaya, Gowa. My summary of the epic follows that provided by Leonard Andaya supplemented by reference to a translation of the original manuscript prepared for me by Muhammad Nasir of Ara (Andaya 1979: 365–366; see also Cummings 2001; for a Bugis version, see Siradjuddin Bantang, 1988).

The Sinrili' Tallumbatua *(Jongaya, Gowa, 1936)*

1. Introduction: The king of Gowa calls together his vassals and his Royal Council and commands them to build stone walls around the palace so high he will be invincible. He summons his sooth-sayer, Karaeng Botolempangang, and asks for his reassurance that he is indeed invincible. Botolempangang says that he is not: he can still be defeated, not by a foreign enemy but by a member of his own family who is still in the womb.

2. The Search: The king follows the suggestion of the *Tumailalang Toa*, Senior Minister of the Interior, suggests that he kill all pregnant women in the kingdom. But when Botolempangang is again summoned three months later, he says his enemy is still alive. The same recommendation to kill all youths of the relevant ages are carried out after seven months, ten months, fifteen months, twenty four months, seven years, thirteen years, and fifteen years. Each time, Botolempangang reports that the king's enemy is still alive and well. Finally, Botolempangang suggests that the king hold a *raga* match to discover the enemy's identity. All the great lords compete, but none emerges as clearly superior. Finally the king's own son and heir, Andi Patunru, is persuaded to join the match. He easily outplays all the others. When he kicks the ball through a palace window, Botolempangang identifies him as the one who will cause his father's downfall.

3. The Flight: The king orders all his nobles to attack his son. Andi Patunru retreats with his half brother and flees to Marusu', where the locals defend them and enable them to escape. Andi Patunru then begins a long journey in search of a ruler willing to help him avenge himself on his father. He stays with the kings of Labbakkang in Pangkaje'ne, of Sidenreng, of Bone and of Bantaeng, all of whom offer him shelter but are afraid to openly antagonize the invincible king of Gowa. Many offer him their daughter in marriage and their kingdom if he will give up his quest to revenge himself on his father. He refuses all such offers. Finally he arrives in Lemo Lemo. The *gallarrang* accompanies

him to Bira to get a boat for Buton. The Karaeng of Bira provides him with the three boats of the title, and he sets sail for Buton. The Karaeng of Buton offers him a daughter, but refuses otherwise to help him. He stays there for three years, by which time news of his whereabouts reaches the King of Gowa. The king sends 42 boats each holding 150 warriors to attack Buton. Andi Patunru is hidden in an old well and the king of Buton swears a fearful oath that he is not there. When they fail to find him, they return to Gowa. Andi' Patunru sails on to Bima, then to Sumbawa, Bali, Buleleng, Solo, and finally to the Karaeng Belanda, the king of the Hollanders, who writes him a note to take to the [Governor] General in Batavia.

4. The War: The General agrees to help in return for Andi' Patunru's help in his campaign against Pariaman in Minangkabau. A long series of wars between the Dutch and Andi' Patunru and the Kingdom of Gowa begins. During negotiations, Andi' Patunru says he will call off the attack if he is allowed to take away his birth mother, his wet nurse, his nanny, and his foster mother. The king's followers tell him to fight on, however. After withstanding a long siege, Gowa is finally forced to sign a treaty when its food supplies run low. Andi' Patunru is restored as the Crown Prince of Gowa and gains the Bugis title of Arung Palakka from his mother. "The story ends with the Tunisombaya and the governor-general in Batavia affirming their friendship and brotherhood." (Condensed from Manuscript 182 held in the Institute of Culture in Ujung Pandang)

In his analysis of this *Sinrili'*, Andaya stressed the role played by the Makassar concepts of *siri'*, shame; *pacce*, fellow feeling and compassion; *sare*, fate; and *taka'dere'*, that which is predestined by God. Andi' Patunru is seen as motivated by an entirely appropriate need to avenge himself for the insult of his father's attack during the raga match, and out of longing for his four mothers. His attempt to seize control of his fate, *sare*, is seen as culturally appropriate because he is acting within the norms of immemorial custom, *ada'*. By contrast, the king is seen as courting God's punishment through his attempt to know that which is predestined by God and through his arrogant attempt to become the most powerful ruler in the world. Thus while the fall of the empire might have been experienced as a great catastrophe by the high Makassar nobles who were forced into exile in its wake, ordinary Makassar villagers experienced it as an act of Divine retribution and an opportunity to reassert of customary law (Andaya 1979).

I would like to suggest some further layers of meaning in the *Sinrili'*. First, the conflation of the royal houses of Gowa and Bone that occurs in the *Sinrili'* reflects a fusion that actually occurred later, in the course of the eighteenth century. The two houses began to intermarry during the lifetime of Arung Palakka. The kings of Gowa and Tallo' lost all legitimacy after the death of Shafi al-Din in 1760, so that when the regalia of Gowa ended up in the hands of Sultan Ahmad al-Salih of Bone, he was able to attract the support of the Mountain Makassar and lay claim to the throne of Gowa.

Second, the *Sinrili'* remains faithful to the deepest emotional truths of Arung Palakka's life and that of other ambitious noble men of middle rank, while altering certain details. He was born not in Gowa, but in the Bugis kingdom of Soppeng. His father was not the king of Gowa, but a minor Bugis noble. But, as stated in the *Sinrili'*, he did experience a life of humiliation at the court of Gowa in his youth. He arrived at the royal court of Gowa as a hostage at the age of eleven, where he was incorporated into the entourage of Gowa's Chief Minister Karaeng Patingngaloang of Tallo'. Furthermore, his mother was the granddaughter of the first sultan of Bone, and so the *Sinrili'* is correct in deriving his claim to kingship from his mother and in distancing him from his father.

The *Sinrili'* also expresses some of the deeper emotional truths of Malay and Indonesian family relationships. Janet Carsten has contrasted the tense and competitive nature of the relationship between fathers and sons, especially in high-status families, with the warm intimacy of the mother–child bond (Carsten 1997: 74). The relationship between Andi' Patunru and his father is a fact established at the moment of conception and is never nurtured thereafter. Their relationship is purely formal and ascribed: there is no mention of any emotional tie between them. The epitome of the cold and distant father, the king declares war on the sons of the whole nation, from their time in the womb right through their childhood. No matter how hard the king tries to extinguish the ties that bind his fate to that of his son, however, he cannot do so.

By contrast, the relationship between Andi' Patunru and his four mothers is almost purely substantive and achieved: one mother bore him in the womb, one suckled him, one looked after him as a toddler, and one fostered him as a child. This is kinship as socially constructed, as the product of a continuing moral relationship. During his travels, he refuses all opportunities to form a relationship with a new woman as a husband rather than as a son. This would fix him in a new, subordinate center away from Gowa. It is as much his love for the four women who played different aspects of the role of mother for him that draws him back to the center as it is the need to avenge himself on his father.

The *Sinrili'* ends with the return of Andi' Patunru, the Conquering Prince, to the royal center. There he takes the title of Arung Palakka from his Bugis birth mother, not from his Makassar father. Ascriptive rank and kinship is repudiated, achieved rank and kinship affirmed. Finally, the story of Andi' Patunru represents the values of popular as opposed to court Islam. The model of the Prophet Muhammad as portrayed in the *Barasanji* clearly provides one source for the *Sinrili's* treatment of Andi Patunru's childhood. Both the Prophet and Andi' Patunru were born into a family of high rank but lacked the protection of a powerful father. Their childhoods were overseen by a series of four women to whom they were linked by intimate emotional ties: a birth mother, two wet nurses, and a foster mother. Finally, they both had to flee from their homes (in Mecca/Gowa) and find external allies (in Medina/Batavia) before they could return to take their rightful places as just rulers.

As we saw in chapter 4, Arung Palakka tried to incorporate the royal house of Gowa into that of Bone by marrying his heir, Idris, to the daughter of Sultan Abd al-Jalil of Gowa. His attempt failed when Siraj al-Din deposed their son, Ismail, and claimed the throne of Gowa for himself. Ironically, it was Arung Palakka who was physically incorporated into the royal house of Gowa when he was buried next to his Gowanese wife, Daeng Talele, near the graveyard of the kings of Gowa. Their joint tomb is venerated to this day by Bugis and Makassar alike (Andaya 1979: 371).

By the nineteenth century, Arung Palakka had thus come to be regarded not as a traitor who collaborated with the Dutch infidel to bring down the sultanate of Gowa, but as equivalent to a Sufi *shaikh* whose achievements in life demonstrated his possession of divine grace, and who continued to be a conduit for divine blessings in death. This version of the past is at odds with the official narrative of the region promoted by the modern nation-state, for whom Sultan Hasan al-Din of Gowa was the true national hero. It encapsulates the popular view of prophecy and righteous kingship that developed during the nineteenth century in Bone's sphere of influence along the Makassar coast. In the course of the *Sinrili'*, Andi' Patunru wins the sympathy of the rulers of all the kingdoms in South Sulawesi before he acquires the VOC as an ally.

Ordinary villagers were well aware that the power of the sultans of Bone had been based on alliances with European powers since the seventeenth century. Unlike the directly ruled territories of the VOC, however, ordinary people in indirectly ruled kingdoms were not subjected to the indignity of submitting to the whims of non-Muslim political superiors. It was possible to deny that alliances with Europeans had led to a fatal compromise of religious principles. These alliances were legitimated in pragmatic terms, as necessary to bring down the greater evil of tyrannical kings. By the late nineteenth century, the true origins of the Bone's power in an alliance with the VOC to bring down the empire of Gowa could be reinterpreted in a different religious idiom. In the *Epic of the Three Boats*, the king is represented as losing his authority through his arrogant assertion of invincibility. The legitimacy of the pretender to the throne is established by modeling his life on that of the Prophet, viewed as a humble orphan who denounced the pretensions of all worldly rulers.

Conclusion: Charismatic Orphans and Androgynous Tombs

Popular narratives such as the *Riwayat Shaikh Yusuf*, the *Sinrili' Datu Museng*, and the *Sinrili' Tallumbatua* represent different responses to the collapse of royal power in the face of Dutch mercantilism. In each case, the hero is a humble figure who begins life with no help from his father. The biological fathers of Yusuf and Datu Museng are unknown. Shaikh Yusuf is repudiated by his stepfather, Datu Museng by his wife's father and Andi' Patunru by his real father. Yusuf and Datu Museng leave the lands of their births after being rejected by the king as too low in rank to marry his daughter. They acquire high rank by traveling away from regional centers of social and political hierarchy to the cosmopolitan center of the religious hierarchy. Andi' Patunru leaves a seemingly invincible imperial center and travels through a series of subordinate royal centers until he arrives at an entirely novel political center. His alliance with this new source of political power allows him to return home, marry the princess, and assert his political authority during his lifetime.

All three heroes are sustained in their exile by the love of the women who serve as their sisters, mothers, nurses, and wives. The charismatic power the male heroes acquire in their wanderings must be linked to the ascribed power of hereditary local princesses. In all three cases, the heroes acquire their greatest power only after they have united with their royal wives in an androgynous tomb, opening up a source of endless blessing and fertility for those who come after them.

The most powerful tombs are those of Shaikh Yusuf and his wife, Daeng Nisanga, the daughter of a sultan of Gowa. Yusuf is a *shaikh* of pan-Indian Ocean importance. He is venerated by Muslims from all over Indonesia and beyond. The dual tombs of Datu Museng and

Maipa Deapati are situated near the old VOC Fort Rotterdam and are venerated by all Makassar as the source of blessings which, unlike the royal ancestor cults, transcend the vertical boundaries between noble houses and the horizontal boundaries between the ranks of noble, commoner, and slave. The dual tombs of Arung Palakka and Daeng Talele are situated near those of the other kings of Gowa and Tallo' and are venerated by both Bugis and Makassar. The androgynous tomb becomes a source of blessings for the groups that claim metaphorical descent from the couple they contain: the entire *umma* in the case of Yusuf, the Makassar subjects of the VOC in the case of Datu Museng, the Bugis-Makassar people in the case of Andi' Patunru.

Despite their similarities, the three stories also exhibit a range of alternative responses to the reality of Dutch military superiority. The *Riwayat* of Shaikh Yusuf portrays its hero as helping the sultans of Banten and Gowa to enforce the religious law within their kingdoms and to defend Islam against external attack by infidels. When this no longer proves possible, Yusuf withdraws into otherworldly mysticism and spreads his teaching throughout the Islamic world. He achieves a moral victory by converting the Dutch governor of Banten to Islam. The epic of Datu Museng represents a response to overwhelming Dutch power that questions the value of all worldly struggles in relation to the reward of the afterlife. It transforms the popular memory of a losing struggle against evil overlords into a mystical narrative in which all earthly existence is portrayed as a tragic separation from the Beloved, and in which death can be welcomed as a return to the Unity of God's Being. Datu Museng achieves his moral victory by consigning the Dutch governor of Ujung Pandang to hell while he and Maipa Deapati ascend to heaven.

The Epic of the Three Boats represents a third kind of response to Dutch power, one that treats it as a fact of life with no obvious moral import. As the most powerful ruler in the archipelago, the governor general of the VOC is the only one with the military capability to help Andi' Patunru in his righteous quest to avenge himself on the clearly immoral emperor of Gowa. Andi' Patunru achieves his moral victory over the Dutch governor general in Batavia by transforming him from a potential enemy into a close ally.

A remarkable feat of symbolic work is thus accomplished by these three epics. On one level, they dramatize the superiority of the charismatic power of an orphan over the traditional authority of a hereditary king. In the *Riwayat Shaikh Yusuf*, the sultan of Gowa is glad to drink the water from Yusuf's tomb after having rejected him as a suitor for his daughter. In the *Sinrili' Datu Museng*, the sultan of Sumbawa is defeated in battle by Datu Museng after rejecting as a suitor for his daughter. In the *Sinrili' Tallumbatua*, the sultan of Gowa is defeated in battle by Andi' Patunru after having tried to murder him several times on the advice of his soothsayer. In all three cases, the heroes gain their charismatic power during a period in exile from their homeland and return in triumph after an absence of several years. This aspect of the narrative is clearly modeled on the biography of the Prophet Muhammad.

On a second level, these epics encode a set of models that allow people to respond to alien political overlords in a variety of religiously sanctioned ways. The example of Shaikh Yusuf allows actors to contemplate armed resistance to Europeans as a religiously sanctioned form of *jihad* when it is practical, and withdrawal into mystical contemplation as permissible when it is not. The example of Datu Museng allows actors who face unbearable humiliation at the hands of a totally corrupt overlord to welcome death in single combat as a religiously sanctioned form of martyrdom. The example of Andi' Patunru allows actors to treat even a non-Islamic political overlord such as the VOC (or even the secular government of President Suharto in the 1980s) as a morally neutral entity that Muslims can put to religiously sanctioned purposes such as removing tyrannical kings.

Finally, these epics manifest the beginnings of a supra-local national identity. In each case, local Muslims who belong to a variety of ethno-linguistic groups and who are subject to many different rulers identify themselves as sharing similar social, political, and religious values in opposition to those of the Dutch. But it was only after the Dutch power had been consolidated in the form of a bureaucratic colonial state that a true national identity would begin to emerge.

Chapter 6

Cosmopolitan Piety and the Late Colonial State, 1850–1950

On the way to the beach, I passed Panre Abeng's old graveyard. The family plot now bears the following inscription: "Makam Tenri Abeng Datu ri Watu, Petta Matinroe ri Appa Parang. Datang di Ara Thn. 1890 (Battan Thn. 1890) Hal. 24" [Lord Abeng, the Datu of Watu [in Luwu'], The Lord who Sleeps in Appa Parang [the name of the field in which he is buried]. Arrived in Ara in the Year 1890 (Batten 1938) p. 24]. Inside are the names "Nusi, H. Gama, Tajuddin, PANRE ABENG." The P, A and E have been plastered over. So it would appear the campaign to retroject Abeng's noble origins continues: he is now not only a Datu but has a death epithet.

(Fieldnotes, May 24, 2000)

Death epithets are only awarded to the highest nobility of the greatest traditional polities. The annotation refers to a colonial *controleur*'s annual report on Ara. The old stone inscription had been defaced so that Panre, a title referring to in this case to his technical skill as a goldsmith, could be rewritten as Tenri, a noble title in Selayar. Clearly, a project to revise the history of this ancestor was in progress. This project had been initiated by a great grandson of Panre Abeng called Haji Basri. It represents a classic example of retelling an historical narrative so that the opening episode, Abeng's birth, would foreshadow the closing episode, Haji Basri's own social position, in an appropriate way.

In this chapter, I analyze a variety of conflicting oral narratives I collected in Ara about Panre Abeng and his descendents. To make sense of these narratives, I relate them to certain policies of the late colonial state in Indonesia and to the development of a pan-Islamic consciousness. The introduction of steamships created pockets of prosperity in the outer islands of Indonesia by enabling the colonial state to finally clear the seas of pirates, and by greatly lowering the cost of transporting tropical produce to Europe. Steamships also greatly lowered the cost of transporting people to Mecca, so that growing numbers of villagers from prosperous areas such as Selayar were able to perform the *hajj*. They brought home with them a pan-Islamic consciousness that regarded many local customs as idolatrous, including those that underpinned the traditional authority of the noble class. Under the guidance of the Dutch orientalist Snouck Hurgronje, the late colonial state attempted to neutralize the destabilizing effects of this consciousness by enforcing a strict separation between religious and political activism. As long as the *hajjis* confined themselves to purifying local symbolic life of non-Islamic practices and refrained from criticizing the government, they could even be seen as allies in the effort to modernize native society.

When Panre Abeng moved from Selayar to Ara in 1890, he brought with him the puritanical critique of local customs that had been developed there by local *hajjis*. His son, Haji Gama, came to power in 1915 by ingratiating himself with the Dutch authorities and by propagating his father's religious views. He maintained a firm grip on the village until 1949 by repressing the cult of the royal ancestors that had legitimated the traditional authority of the hereditary rulers of the village. He legitimated his own rule exclusively through the charismatic authority he derived from his mastery of esoteric knowledge, *ilmu*, and from his sponsorship of weekly recitations of the *Barasanji* and processions to the tomb of the local *shaikh*. Haji Gama knew enough of the Arabic script to recite the Koran, but relied on a scribe to read and write government documents written in Malay using the Roman script. He ended his career by accepting the colonial government's offer to pay for his performance of the *hajj* in 1949, when the war of national liberation was at its height. Many of his children and grand-children went on to achieve high bureaucratic office in the Republic of Indonesia.

The Late Colonial State, 1850–1910

The introduction of steamships in the 1850s finally allowed the colonial powers to bring piracy under control throughout Southeast Asia (Warren 1981). Dutch steamships were particularly active off the coast of South Sulawesi between 1850 and 1853 (Berigten 1855: 15–24; Kniphorst 1876). The introduction of steamships also cut the costs of transporting bulk tropical produce to Europe dramatically. As cargo rates fell and as demand for tropical produce such as copra, sugar, and rubber rose in the late nineteenth century, many areas in the outer islands of Indonesia experienced a new kind of prosperity. By 1885 private exports from the East Indies as a whole were ten times those of the government (Ricklefs 1981). Some of the new wealth was spent on the *hajj*, which also became cheaper with the introduction of passenger steamships. With the opening of the Suez Canal in 1869, the steamship route to Europe passed through Red Sea, and pilgrims traveling in steerage could easily be put off and taken on at the port of Jeddah along the way. This led to further increase in the number of *hajjis*.

A persistent anxiety arose among Dutch officials that returning *hajjis* might spread subversive pan-Islamic ideas when they returned to their villages. A number of measures were introduced to keep tabs on them. An ordinance issued in 1859 required every *hajji* to obtain a certificate from his regent stating that he had sufficient means to perform the *hajj* and to maintain the family members he had left behind. Upon his return, every *hajji* had to pass an examination proving he had actually visited Mecca. Only then was he allowed to use the title and dress of a *hajji* (Vredenbregt 1962: 100–103).

Prosperity and Puritanism in Selayar, 1860–1884

The island of Selayar in South Sulawesi was one of the first areas to benefit from the suppression of piracy and the introduction of cheap shipping. The farmers of Selayar took to the intensive production of copra. By 1860, Selayar was responsible for over 70 percent of the coconuts exported from South Sulawesi. By the 1880s, many elite families in Batangmata, Bonea, and Bontobangun in Selayar owned between 10,000 and 20,000 coconut palms. They used much of their new wealth to finance pilgrimages to Mecca. In 1863 alone, forty people made the *hajj* from Selayar. By the late nineteenth century, a disproportionate number of *hajjis* in South Sulawesi were coming from Selayar each year, and within Selayar they were largely coming from one area. Of the 179 *hajjis* living in Selayar in 1879, 111 were from the regency of Batangmata. Forty-two of these 179 *hajjis* were women (Heersink 1995).

Many of these *hajjis* made it their business to enforce a puritanical form of Islam. They took particular exception to the performance of *pakarena* dances. H.E.D. Engelhard, a Dutch official who was posted to Selayar in 1877, published two long accounts of the island in which he lamented the loss of the folk traditions opposed by the *hajjis*, especially the "Selayarese national dances" that had already died out on the

mainland. Those performed by young noble women commemorated
Selayar's ancient allegiance to the kings of Gowa, while the martial
songs and dances performed by young men commemorated Selayar's
fealty to the VOC during the eighteenth century (Engelhard 1884a:
310–313).

Direct Rule in Bira, 1860–1910

The colonial government began its first serious attempt to exert direct
authority over the villages of the Bira peninsula following the Second
Bone War of 1859–1860. In May 1863, authority was granted to the
governor to "gradually decrease the number of regencies and
galarangships" in the old Division of Boelecomba and Bonthain
(Goedhart [1920] 1933: 140). A new unit called the Eastern Districts
was created that included Bantaeng, Bulukumba, and the Bira peninsula,
as well as areas such as Kajang and Sinjai that had just been wrested
from Bone's control (see map 6.1).

Matthes made his second tour through the Bira peninsula in 1864.
He submitted a detailed account of conditions in the area to his old
friend, J.A. Bakkers, who was now governor of Celebes. One month
later, the government began the process of rationalizing and consolidating the units of local government in the area. Ara was absorbed by
Lemo Lemo, and Bonto Tanga was absorbed by Tiro. Tiro absorbed
Tanaberu in 1865 and Batang in 1867. In 1869, Bira incorporated
the already combined villages of Lemo-Lemo and Ara. In the same
year, Bira was included in the new *onderafdeeling*, subdivision, of
Kajang and placed under the authority of its Dutch *controleur*.

At this point, the Subdivision of Kajang was composed of the
Regencies of Kajang proper, Wero, Tiro, and Bira. Finally, Bira incorporated the enlarged Tiro in 1871 upon the death of its Regent,
Lewai Daeng Matana (Goedhart [1920] 1933: 141). The *karaeng* of
Bira who had been in office since 1849, Baso Daeng Raja, now
became the regent of a territory that included seven formerly independent realms, all of whose rulers were demoted to the rank of
gallarrang.

This consolidation provoked stubborn resistance on the part of the
old ruling families in each of the seven realms. The regent of Bira
reported to the *controleur*, a Dutch civil servant who was usually a
young man at the very beginning of his career and who usually
remained in place for only two years. This enabled the regents and
gallarrangs to run local affairs pretty much as they pleased. Another
problem was that the regent and his deputy, the *sulewatang*, were not

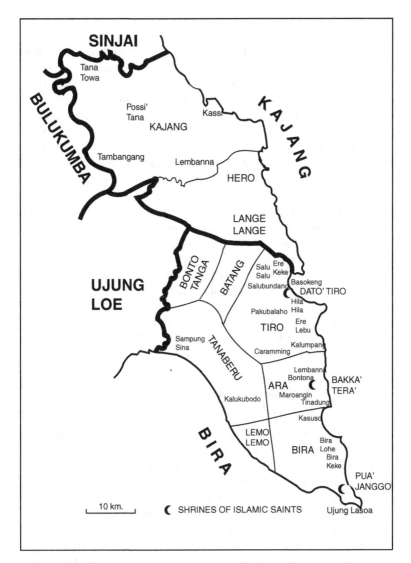

Map 6.1   The Regencies of Bira and Kajang, 1874–1920

paid a salary. They were expected to support themselves by levying a variety of "customary payments," *pangadakang*, for performing their official duties at life-cycle rituals and upon the completion of boats in the boatyards of Ara and Lemo Lemo. But these payments traditionally belonged only to the hereditary officials in the villages and they refused to surrender their claim on them to the regent. This led to endless quarrels between the regent and his deputy, and between the two of them and the village chiefs (Goedhart [1901] 1933: 224).

The political offices that were entitled to *pangadakang* payments in nineteenth-century Ara included that of the *gallarrang*, or lord of the whole village; the *kepala*, or the chiefs of the settlements of Ara proper, Lembanna and Tinadung; and the *sariang*, or the emissaries of the rulers. The religious offices entitled to *pangadakang* payments included that of the *kali*, or chief religious official of the village; the *katte* (from Arabic *kha'tib*, preacher), or those in charge of delivering the Friday sermon at each mosque; and the *bidal* (from Arabic *Bilal*, the name of the first Muslim to call the faithful to prayer, i.e., the *muezzin*), or those in charge of issuing the call to prayer at each mosque.

In 1901, Goedhart wrote an extensive report on the failure of the whole attempt to rationalize local government in the area. He explained this mostly as the result of the aversion to paying *pangadakang* to the regent and *sulewatang* of Bira, since they had no traditional right to it. He advocated placing them on regular salaries financed by a uniform head tax. He thought that the *pangadakang* should be maintained for those holding political and religious offices at the village level since the people paid them quite willingly, but that the government should codify these payments (Goedhart [1901] 1933: 226). Goedhart's advice on this matter was not followed until 1920.

The Ethical Policy and the Forward Movement, 1901–1910

The ambitions of the late colonial state to transform the societies of Indonesia took another step forward in 1901, when a coalition of right-wing and religious parties came to power in the Netherlands. This coalition adopted the moralistic stance toward the colonies propounded by van Deventer, who held that the Netherlands owed the East Indies a debt of honor for past abuses. Queen Wilhelmina officially acknowledged the nation's "ethical obligation and moral responsibility to the peoples of the East Indies" and appointed a commission to inquire into local welfare in Java (van Niel 1984: 32). The ethical policy was primarily implemented by Alexander Idenburg during his tenure as minister of the colonies (r. 1902–1905, 1908–1909, 1918–1919) and as governor general of the Netherlands East Indies (r. 1909–1916).

Since such measures could only be undertaken in areas that were firmly under government control, the ethical policy was accompanied by the "Forward Movement," in which the last autonomous areas in the Netherlands East Indies were absorbed into the colonial state by military force. The official adviser on Islamic and native affairs from 1891 to 1906 was Snouck Hurgronje. He held that so long as Islamic leaders steered clear of politics, the government should stay away from religion. Thus the examination for performing the *hajj* was abolished in 1902 and the certificate proving one's financial means was eliminated in 1905.

But when Islamic leaders did involve themselves in politics, Snouck Hurgonje believed that the government had to punish them. The Acehnese began fighting occupation by the Dutch in 1873 and by 1881 the resistance was largely led by the *ulama*. Little progress was made until 1898 when J.B. van Heutsz was appointed governor of Aceh and began to follow Snouck Hurgronje's advice. "He maintained that nothing could be done to appease the fanatical resistance of the *ulamas*, so they should be utterly crushed and reliance placed upon the *uleebalang* (seen as the *adat* or 'secular' chiefs)" (Ricklefs 1981: 137). In 1903, Sultan Tuanku Daud Syah surrendered.

The conclusion of the Aceh war freed sufficient military resources to complete the "pacification" of the rest of the archipelago. When van Heutsz was appointed governor general in 1905, he ordered the imposition of direct rule over all of South Sulawesi, beginning with Bone. Relations between Bone and the Dutch had been relatively smooth during the reigns of Sultan Ahmad Singkarru (r. 1860–1871); his daughter, I Banri (r. 1871–1895); and her brother, La Pawawoi (r. 1895–1905; IJzereef 1987). But by 1905, the government was looking for an excuse to invade the kingdom. La Pawawoi gave them one when he refused to impose certain import and export duties or to pay an indemnity he owed the government. La Pawawoi was captured and exiled to Java, and the kingdom of Bone was transformed into a district under the direct administration of a Dutch resident.

The Dutch then put pressure on Sultan Husain of Gowa (r. 1895–1906) to submit all of his actions for approval. The sultan refused and fled to Enrekang, where he was killed in battle in 1906. His brother and son were captured and exiled to Java, and Gowa was also transformed into a district under a Dutch resident. By 1910, three centuries of Islamic kingship in South Sulawesi were at an end (Harvey 1974: 53–54).

Puritanical Islam in Ara, 1890–1930

The puritanical attitudes of the Selayar *hajjis* were brought to Bira and Ara by migrants in the 1880s and 1890s. Among the most influential was a man called Panre Abeng. Although he had acquired a somewhat legendary status by 1988, nobody seemed to know much about his precise origins. This was partly because he died before any of his grandchildren were born and all the stories about him were secondhand. Very little could be discovered about Abeng's parentage even in the 1920s, however, when his son was trying to legitimate his appointment by the Dutch as *gallarrang* of Ara. Most of my informants agreed that Panre Abeng had moved from Selayar to Bira at the height of the copra boom in about 1880. *Panre* is a title meaning "expert" that is applied to those with a technical skill such as working with gold. Since goldsmiths often worked as dentists, it was assumed by many that Panre Abeng had traveled from town to town repairing people's teeth.

After moving to Bira, Abeng married a woman from Ara called Nusi. On one genealogy I saw, Nusi was shown as the first cousin of Ganna, the *anrong tau* of Lembanna, or chief of the lower settlement in Ara. This would indicate that Abeng had managed to make a fairly good marriage in terms of the local social hierarchy, perhaps because of the wealth he brought with him from Selayar and Bira. Abeng and Nusi had a son called Gama Daeng Samanna in 1884 and moved to Ara in 1890. Abeng's religious principles were embedded in the names he gave his children. Gama is derived from the Malay word for religion, *agama* (which is itself derived from Sanskrit). Another son was called Adam, who is regarded as the first prophet in Islam. His daughter was called Sadaria, from the Arabic word for penitence. Such names were highly unusual in the nineteenth century, when almost all personal names were still drawn from Makassar sources. Abeng was such a ritual purist that he regarded all of the existing graveyards as having been desecrated by ancestor worship, a form of *shirk*, or idolatry. When he died in 1910, his family opened an entirely new graveyard for him. Only his descendents were buried there until the 1950s, when it was made available to others who shared the family's principles.

In Ara, the story of the first half of the twentieth century is the story of the rise to power of Panre Abeng's eldest son, Gama Daeng Samana. The following account of Gama's early years comes mostly from Patoppoi, the only one of Gama's sons who was still living in Ara in 1988. Gama grew up in the lower settlement of Lembanna, which had a rather riotous reputation at the time. It was a center for the drinking of palm wine and for gambling. The family of the *anrong tau* of Lembanna, Ganna, was also deeply involved in the cult surrounding the royal ancestor spirit, Karaeng Mamampang. Karaeng Mamampang regularly spoke to his descendents through a spirit medium called the *karihatang* (from Sanskrit *dewata*, god). The *karihatang* until 1910 was a man called Raduna. His daughter, To Ebang, married Ganna's brother,

Mangga. To Ebang inherited the royal ancestor spirits from her father in 1910, and served as the *karihatang* until her death in 1961. Ganna's daughter, Anni, served as To Ebang's *jurubasa*, interpreter, for most of that period (see figure 6.1).

Gama was brought up on Abeng's "fanatic" Islamic principles in this rather pagan environment. He learned to read enough Arabic to recite the Koran. He also acquired a great deal of *ilmu*, esoteric knowledge, by fasting and meditating on top of powerful tombs on Thursday nights. Supernatural beings such as *jinn* and the spirits of the village *shaikhs*, Bakka' Tera', would then appear to him and teach him *mantera*, powerful incantations that gave him the ability to facilitate childbirth, to cure disease, and to remain impervious to penetration by weapons in battle. Acquiring knowledge in this way requires a great deal of courage. Abdul Hakim was fond of telling a story about another man who had tried several times to gain *ilmu* from Bakka' Tera', but was frightened away every time before he could ask a question. Finally he tied himself with a stout rope to a heavy gravestone so he could not run away. Nevertheless, he got up when the first apparition came, and ran all the way home with the tombstone tied to his back. It later took four men to return it to the cemetery.

About the time of Abeng's death in 1910, Gama's strength and courage came to the attention of the regent of Bira, Andi' Mulia (r. 1900–1914, 1931–1942). A notorious thief, Daeng Patanra, had been rustling cattle on the plateau above the village of Ara. As he was slipping back into the village one night, Gama leapt out, grabbed him by the thumb, and marched him all the way down to Bira for judgment. When Gama returned to Ara after this incident, he had a dream in which he was holding the sun and the moon wrapped up in his sarong. When he awoke, he was still holding his sarong in the shape of a bundle. He told his friend Daeng Majambeang about it, who commented that he was surely destined for great things. Gama made his ambition for high office plain soon afterward when the *gallarrang* of Ara, Daeng Makkilo, sent Gama to fetch his hat for a formal dinner he was attending. Gama returned wearing the *gallarrang's* hat on his own head. When Ganna retired from his position as *anrong tau* of Lembanna, Daeng Makkilo appointed Gama to replace him.

In 1913, Daeng Makkilo suddenly died in a cholera epidemic. His half brother, Daeng Pagalla, took over as acting *gallarrang*. In 1914, Andi' Mulia stepped down as regent of Bira to go on the *hajj* and was replaced by his assistant, Uda Daeng Patunru. Daeng Patunru's appointed marked a significant shift in Bira. Daeng Patunru was sympathetic to the puritanical ideas coming out of Selayar and had

146

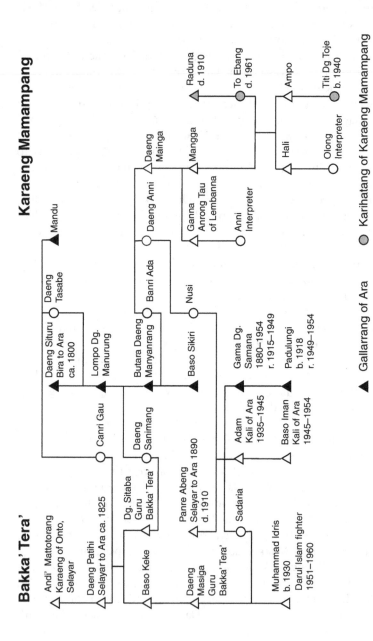

Figure 6.1 The Cults of Bakka' Tera' and Karaeng Mamampang

been friendly with Gama's father when he was living in Bira during the 1880s. Daeng Patunru appointed Ongke, a reformer who shared their views, as the *kali* of Bira. One of Daeng Patunru's sons, Andi' Baso, married Ongke's daughter and later succeeded his father-in-law as *kali* of Bira. Daeng Patunru appointed another one of his sons, Nape Daeng Mati'no, as his *sulewatang*, official assistant.

As we have seen, the seven villages of the Bira peninsula had only been incorporated into one regency in the years following the Second Bone War, and the noble families in these villages had always resisted the authority of the regent. Daeng Mati'no was given the task of enforcing his father's wishes in the outlying villages. Among his strongest supporters in Ara was Pantang Daeng Malaja. Daeng Patunru had virtually adopted Daeng Malaja and left him one hundred coconut trees when he died. Daeng Patunru's son, Daeng Mati'no, later acted as the patron of Daeng Malaja's children. They spent much of their youth in Daeng Mati'no's house and he helped all of them to get married. Daeng Malaja's three sons, Pasohuki, Muhammad Nasir, and Haji Arifin, and his daughter, Saborang, were among my best informants in 1988 and 1989. According to Pasohuki, Daeng Malaja was noted for his bravery and toughness. He served as Gama's *Palappi' Barambang* "He Whose Breast Shields," or bodyguard. He served as Daeng Mati'no's chief deputy and tax collector in the regency of Bira between 1904 and 1920. His children maintained that it was his very fierceness that enabled him to protect the people of Ara from the excesses of Daeng Mati'no.

The colonial government decided to hold an election to select a permanent *gallarrang* of Ara in 1915. Daeng Mati'no decided to use this opportunity to break the power of the noble families there by deposing the acting *gallarrang*, Daeng Pagalla, and by appointing his own protégé, Gama. Daeng Pagalla's nephew, Daeng Majannang, told me that Daeng Mati'no knew Gama would lose if the election were held in Ara, so he moved it to Hila Hila in Tiro. Of the fifty men who went there to vote, forty-eight voted for Daeng Pagalla and only two voted for Gama. When the *karaeng* of Tiro, Tonang Daeng Patiho, declared Daeng Pagalla the winner, Daeng Mati'no intervened and declared that as *sulewatang* his vote was equal to fifty ordinary votes from Ara, and that Gama was therefore the winner. Daeng Pasau, the *kepala desa* while I was in the field, agreed with this account generally, but said there had been seventy votes for Daeng Pagalla and fifteen for Gama.

Gama's son, Patoppoi, gave a different account. He claimed that Gama had been legitimately chosen not by a popular vote, but by the *hadat* of Ara. The *hadat* was an electoral council found everywhere in South Sulawesi. In royal foundation myths, the first *hadat* was composed of a group of local men who encountered the first royal ancestor after he or she had descended from the heavens onto rock. These myths end with the members of the *hadat* swearing an oath to obey the *Tomanurung*, The One Who Has Descended, so long as the *Tomanurung* maintains the social harmony and natural fertility of the realm. This mythical template was reenacted every time a new ruler was installed. The *hadat* chose a new ruler from among all the eligible heirs of the last ruler and swore him in while he stood on the *palantikan*, the rock on which the founding royal ancestor had descended. In Ara, the *hadat* was composed of three territorial chiefs (the *anrong tau* of Bontona, Lembanna and Tinadong) and of six religious officials (the *kali* and *imam* of Ara; the *katte* and *bilal* of Bontona; and the *katte* and *bilal* of Lembanna). Even Patoppoi conceded, however, that Gama's appointment was hotly disputed. The hereditary nobles filed a lawsuit claiming Gama was disqualified for office by his low birth. This is when Gama tried to prove that his father had come from a noble family in Selayar, but was unable to do so.

Gama's situation became even more tenuous when his patrons were removed from office in 1920. Daeng Patunru was accused of embezzling government tax money and of torturing people suspected of theft. He was removed from office and exiled for eight months. (Daeng Patunru was still living in Bira when Collins arrived. He refers to him as the "First Old Karaeng.") His son, Daeng Mati'no, was accused of torture and perhaps even of murder. He was removed as *sulewatang* of Bira and exiled to Java for ten years. As Pasohuki told it, Daeng Mati'no was accused of having killed a woman in Tiro by thrusting a burning stick into her vagina. Pasohuki's brothers claimed that he had only threatened to do this to the woman as a means of getting her husband to confess to being a thief. This seemed reasonable behavior to them, because one had to be *keras*, harsh, in those days to maintain the people's respect and to keep order. All three of them said that the true cause of Daeng Mati'no's exile was his nationalist sentiments. He had often stood up to the Dutch authorities, and at one point had the effrontery to bang on the table in front of the governor in Ujung Pandang.

In the wake of this scandal, Goedhart was asked to evaluate the experiment of consolidating the seven villages of the Bira peninsula into one regency. He concluded that the whole project had been doomed to failure because of the way it ran rough shod over local custom. The regent had never exercised any real authority over the district *gallarrangs*, who retained all the real authority because they were chosen from the old *karaeng* and *gallarrang* families. The successive regents of Bira had been put in the untenable position of persuading the village chiefs to follow the government's orders without having the resources or authority to make them do so (Goedhart [1920] 1933: 141–142).

Goedhart recommended that the whole layer of native regents be cut out of the colonial bureaucracy, and that district chiefs be made responsible directly to a Dutch *controleur*. This was in line with a general shift in colonial policy on local government that was then underway. The regency of Bira was broken up again into seven "*gelarangschaps.*" An outsider from Selayar, Baso Daeng Makanyang, was appointed as the ruler of the district of Bira (Collins refers to him as the "Second Old Karaeng," 1937: 140).

In the absence of his patrons in Bira, Gama needed the support of at least a portion of Ara's nobility to remain in office. According to Daeng Pasau, *Controleur* Baljet (1918–1921), persuaded the village *kali*, Baso Daeng Siahing, to break ranks with the other nobles and to support Gama. The lawsuit of the noble families was rejected and Gama was officially confirmed as *gallarrang* of Ara in November, 1921. He was placed on a salary of 30 guilders a month and continued to receive the harvest from the royal *ongko* "ornament" fields at Kaddaro and Buka Lohea, and reduced amounts of *pangadakang*, customary fees (Batten 1938).

Gama named his first few children after the situation he was facing at the time of their birth. His first son was born in 1918 and was called Padulungi, from *dulung*, "war leader," a name that referred to the unremitting opposition to his rule by the noble families. His second son was born in 1920 and was called Palioi, from *lio*, "bristling feathers on a cock," a name that compared the continuing political battle to a cockfight. This son later became regent (*bupati*) of Selayar. His third son was born in 1922 and called Pasauri, from *saura*, "to defeat," because his opponents had just lost their court case. His fourth son was born in 1925 and called Patoppoi, from *toppo*, "to surrender," because his opponents had finally accepted their defeat. Gama had three more sons, but all received normal names.

In about 1925, Daeng Siahing was replaced as *kali* of Ara by Daeng Matanang. Daeng Matanang was descended both from the royal house of Ara and from Haji Ahmad the Bugis, but he was also sympathetic to many modernist critiques of traditional village ritual. He joined Gama in a campaign to strengthen orthodox practices and to eliminate heterodox ones. They replaced the small wooden mosque near the Great Spring with a large stone structure and began to keep track of which men attended Friday prayer services. Those who failed to attend were banned from burial in the Islamic cemetery. They even tried to prevent people from praying for the souls of those they considered impious. They also discouraged the consumption of palm wine, a practice that was still prevalent in the 1920s.

Gama began his attack on heterodox rituals with an attack on rituals in honor of the spirits who lived in large banyan trees. People made regular offerings to these spirits to ensure their benevolence. Gama defied them by personally cutting down their trees. He then turned to the cult of the noble ancestors that served as the symbolic basis of noble rank and power. He climbed right into the attics of noble houses and destroyed any ancestor shrines he found there.

Gama was never able to destroy the shrine of Ara's founding royal ancestor, Karaeng Mamampang, however, or to completely repress his cult (see Gibson 2005: 190–226). Throughout the 1930s, Gama played a game of cat and mouse with Karaeng Mamampang's spirit medium To Ebang. To Ebang would hold her séances in the dead of night in a house far from the village center. Gama would disguise himself in old clothes, sneak up on the house, burst into the room, and put a stop to the séance before they could hide the evidence.

Unwilling by conviction, and unqualified by descent, to participate in the cult of the royal ancestor, Gama became an enthusiastic devotee instead of the cult of the village *shaikh*, Saluku Kati, better known as *Bakka' Tera'*, the Great Belcher. Gama gained access to the cult through his sister, Sadaria, who was married to Cilla Daeng Masiga, the guardian of Bakka' Tera's shrine. Cilla Daeng Masiga was a nephew of Baso Daeng Sitaba, who had served as the guardian of the shrine for much of the nineteenth century. Daeng Sitaba's father, Daeng Patihi, had moved from Selayar to Ara at the beginning of nineteenth century, and he may have recognized a family connection with Gama's father Abeng (see figure 6.1).

According to oral tradition, Bakka' Tera' was a contemporary of Datu Tiro, one of the three Sumatran *shaikhs* that converted the kings of South Sulawesi in the early seventeenth century. The tradition says that Bakka' Tera' acquired his nickname because his good friend, Dato Tiro, could hear his belches five kilometers away in Hila Hila. This tradition places him seventeen generations before the present and is compatible with the claim that he converted Ara to Islam.

According to a genealogical manuscript from Bira, however Bakka' Tera' was born some twenty-one generations before the present, between 1450 and 1475. The manuscript describes him as Gowanese noble who married a daughter or a granddaughter of Ara's founding royal ancestor, Karaeng Mamampang, who was also a noble from Gowa. According to this genealogy, Bakka' Tera' moved to Ara a century before the village converted to Islam and it is highly unlikely that he was originally regarded as an Islamic *shaikh*.

I would suggest, therefore, that before Ara converted to Islam, there were two powerful tombs in the village, each containing the remains of a foreign prince who had married a local princess. This scenario follows a common mythical pattern found along the Makassar coast, where the founding royal couple is composed of a wandering prince from a powerful royal center like Majapahit and a local princess who descends from the heavens into a bamboo internode. It is the fusion of terrestrial male and celestial female power into a single androgynous whole that produces new life (Gibson 2005: 128–138).

When Ara converted to Islam, the tomb of Karaeng Mamampang was preserved as a site for the observance of pre-Islamic royal rituals, while the tomb of Bakka' Tera' was "converted" into that of an Islamic *shaikh*, providing a place where the new Islamic rituals could be performed to tap the new, cosmopolitan source of charismatic power. One can find analogous conversions of the tombs of pre-Islamic royal ancestors into the tombs of Islamic *shaikhs* all over South Sulawesi. These tombs form a province-wide hierarchy of sacred sites, at the apex of which is the shrine of Abd al-Qadir Jilani on top of Mount Bawakaraeng.

The base of this hierarchy is composed of village sites such as the tomb of Bakka' Tera'. The village of Tiro just to the north of Ara contains a sacred spring associated with Datu Tiro, the Sumatran *shaikh*. The village of Bira to the south contains a sacred mountain peak associated with Abd al-Haris, better known as *Pua' Janggo'*, The Bearded Master, an antinomian Sufi mystic who used to meditate there in the late eighteenth century. In the 1980s, Muslim traditionalists continued to make vows and to leave offerings at all three sites when they wished to ask the *shaikhs* to intercede for them.

Patoppoi described for me the rituals of homage, *a'dalle*, that Gama sponsored at the tomb of Bakka' Tera' during the 1930s. At that time the tomb was surrounded by a high stone wall, and only certain people were allowed to enter the enclosure. Every Thursday night, four or five men would go to the tomb, scatter fragrant flowers, recite a series of prayers, and sacrifice a goat. The meat would be taken back to the Gama's house to be cooked and eaten. On holy days such as the Prophet's birthday (*maulid*) and in fulfillment of vows (*nazar*), a more elaborate ritual would be performed. The *Barasanji* would be recited in the sponsor's house, accompanied by an elaborate array of food offerings set out on a *kappara bangkeng*, a silver tray with feet. When the *Barasanji* was over, the offerings would be carried in procession down to the grave. Seven maidens and seven youths would dance around the tomb to the accompaniment of gongs and drums, while the *gallarrang*, the *kali*, and the *katte* officiated over the recital of prayers.

The rituals in honor of Bakka' Tera' were very similar to the ones performed by the *karihatang* in homage to Karaeng Mamampang. Before Gama came to power, the *gallarrangs* of Ara had sponsored rituals in honor of both figures, thereby claiming both traditional and charismatic authority. Gama split the two kinds of ritual and the two forms of authority apart. He used his access to the cult of Bakka' Tera' to legitimate his own charismatic authority, and repressed the cult of Karaeng Mamampang to delegitimate the traditional authority of his noble antagonists.

Dutch Reaction and Islamic Traditionalism, 1926–1950

Gama's attack on traditional authority received tacit support from the colonial government well into the 1920s. But when communist uprisings broke out in Java and Sumatra 1926, and when Sukarno brought together a number of Islamic, Christian and regional groups to form the Indonesian Nationalist Party (PNI) in 1927, the colonial administration turned away from progressive Indonesians and back to the hereditary rulers of the Indies. According to the ethical policy that had been in place until that point, the Indonesian elite would ready for self-rule as soon as they had been taught to value the legal and political institutions of modern Europe. In the 1930s, the ethical policy of Idenberg was replaced by the cultural conservatism of Cornelis van Vollenhoven, who held that native legal institutions could be changed only very slowly. They must first be codified and used as a basis for colonial rule or social anarchy would result. According to his logic, it would be a very long time before Indonesians were ready for independence (van Vollenhoven 1931, 1981).

In 1929, the ethically minded Governor General de Graeff ordered the arrest of Sukarno and other nationalist leaders. In 1931 he was replaced by the openly reactionary Bonifacius de Jonge, a former minister of war and director of Royal Dutch Shell. To complete the picture, "The Minister of Colonies in the Hague from 1933 to 1937 was Hendrikus Colijn, a bitter opponent of Ethical ideas and sometime director of Shell" (Ricklefs 1981: 177). The new conservative policy was soon brought to bear on South Sulawesi, where the Dutch decided to restore the old kingdoms they had just abolished. Government linguists and ethnologists were assigned to codify the rules and ceremonies according to which traditional rulers had been selected and installed, even if they had not been used for more than a hundred years. Royal installation rituals were revived in realms as small as Bira, Ara, and Tanaberu (see Gibson 2005: 181–186).

After returning from his pilgrimage to Mecca in 1917, Andi' Mulia had spent six years traveling around Indonesia, and had returned to Bira in 1923 (Collins 1936: 133–140). A noble from Lemo Lemo called Baso Daeng Makanyang served as *karaeng* of Bira from 1921 until he retired in 1931. Andi' Mulia was persuaded to come out of retirement and take over as *karaeng* again in 1931. During his second term in office Andi' Mulia became a supporter of the Nahdlatul Ulama (NU), a traditionalist Muslim organization that formed in 1926 in opposition to the modernist Muhammadiyah, an organization whose members also tended to be nationalistic (see chapter 7). He replaced the modernist *kali* of Bira, Andi' Baso, with an NU supporter, Haji Andi' Rukha Muhammad Amin. The *karaeng* of Tiro, Tonang Daeng Paoha, was also a strong opponent of the Muhammadiyah until his death in 1937 (Collins 1936: 242). When Daeng Paoha retired with a decoration and a pension after more than forty years of service to the Dutch, he was replaced by one of his pro-Muhammadiyah sons (Collins 1937: 170–171).

The reactionary turn in government policy reached Ara in 1938, when Controleur Batten abolished adult male suffrage and revived the *hadat*, the council of nine office holders who traditionally chose the *gallarrang* of Ara. By this time, Gama's allies had taken control of most of these offices from the traditional nobility. Gama's brother, Adam, served as *kali* of Ara from 1935 until 1945.

The reactionary phase of Dutch rule was briefly interrupted in South Sulawesi by the Japanese occupation, which lasted from February 1942 until August 1945. The Japanese did little to alter the balance of power at the village level. They mostly confirmed the existing local aristocrats in power during this period, including the rulers of Luwu', Gowa, and Bone. They did give some encouragement to Islamic modernists, many of whom viewed the Japanese as a lesser evil than the Dutch. The former *sulewatang* of Bira, Daeng Mati'no, welcomed the Japanese with open arms and was appointed village chief.

Three days before the dropping of the atomic bomb on Hiroshima, the inhabitants of Ara saw a beam of light rising from the tomb of

Bakka' Tera' to the heavens. This presaged the downfall of Japanese power and the birth of an independent Indonesia. Immediately after the surrender of the Japanese, Indonesian nationalists declared their independence from the Netherlands on August 17, 1945.

The first Allied troops reach Makassar on September 21, 1945 and Dutch colonial rule was quickly restored. Fighting began in Java between pro-Republican forces and Dutch colonial troops in the same month and continued throughout 1946 at a low level. Most of the royal houses in South Sulawesi went over to the nationalist side. The Sulawesi delegation to the Indonesian Independence Preparatory Committee (PPKI) in August 1945 included the *karaeng* of Gantarang, a district in Bulukumba. He was deposed and confined to Makassar after fighting broke out. The *karaeng* of Bantaeng was arrested in November 1945 and imprisoned in Makassar.

After heavy fighting, the Dutch were forced to cede Java and Sumatra to the nationalists in November 1946. The Dutch then tried to set up a "United States of Indonesia" in the remainder of the colony. The "State of East Indonesia," *Negara Indonesia Timor* (NIT), was to be a part of this new entity, with its capital in Makassar. All the old kingdoms of South Sulawesi were to be restored. In areas such as Bulukumba, which had been under Dutch rule since 1667, new self-governing entities were to be created on the model of the indirectly ruled kingdoms: "The legal personality so established, described as 'being of the same type as a self-governing land,' has been variously termed as neo-land, neo-self-governing land, fictional land, or neo-administration" (Schiller 1955: 99).

The "neo-land" of Bulukumba was created in January 1947 (Schiller 1955: 363 n.108). It was composed of fourteen adat communities plus the municipality of Bulukumba, following the account Goedhart wrote for the Commission on Adat Law of the traditional communities of the area (Schiller 1955: 105; Goedhart [1920] 1933). The Dutch tried to revive feudal traditions all over the province in hopes that the less educated nobility would support them in opposing the generation of educated nationalists who favored integration with the new Republic.

Resistance to the NIT in South Sulawesi was fierce, even among the high nobles the Dutch were promising to return to power. Andi' Mappanyuki, who had been ceremoniously installed as sultan of Bone in 1931, was arrested on November 8, 1946. He was replaced by Andi' Pabbenteng, a grandson of Sultan La Pawawoi (r. 1895–1905). Between 1946 and 1950, the Dutch replaced between one-quarter and one-half of the traditional rulers in South Sulawesi. Of those removed, about half were killed, 40 percent were imprisoned, and the rest went into agriculture or commerce (Harvey 1974: 158).

In December 1946 the Dutch began an all-out pacification campaign under General Westerling during which thousands were killed or imprisoned. A state of war was declared in the Divisions of Makassar, Bantaeng, Pare Pare, and Mandar. It was lifted in Pare Pare and Mandar in January, 1948, but remained in force in Makassar and Bantaeng until July, 1949. The Republicans claimed that 40,000 were killed, and this figure is still taught to schoolchildren as historical truth. The Dutch admitted only to 2,000. Whatever the true number of victims, the counterinsurgency operation created great bitterness toward the Dutch all over South Sulawesi (Harvey 1974: 128).

> The way in which the pacification campaign was carried out in the countryside also involved Indonesians in the responsibility for the killings. The most usual technique seems to have been to assemble all the villagers in a central area, and to ask them to point out the "extremists" in the group. Those so designated were shot on the spot. If no information was volunteered, several villagers would be chosen at random and shot. (Harvey 1974: 167)

Not surprisingly, these methods left a number of scores to settle throughout the province.

In Ara, Gama was given the title of *karaeng*, the first time a ruler of Ara had borne this title since the village had been incorporated into the empire of Gowa in the sixteenth century. By this time, Gama's position in the village was so secure that he was ready to assert a claim to noble rank openly. He required people to address him as *Opu*, the title of a ruler in Selayar, implying that his father Abeng had in fact been of the noble class. But the only real test of noble rank in South Sulawesi is a successful negotiation of a marriage between a male member of a house with a female member of an indisputably noble house. This is because while men can marry a woman of lower rank, women must always marry a man of equal or higher rank. Gama's next move, undertaken in the mid-1940s, was thus to acquire high-ranking wives for his sons. He achieved this by going outside Ara to the neighboring village of Tanaberu, where he negotiated the marriage of his third son, Patoppoi (b. 1923), to Hasanang. Hasanang was the great granddaughter of Andi' Kinding, a famous princess and spirit medium who was born in about 1835. When she was a child, Andi' Kinding had received a vision from Tanaberu's royal ancestor, To Kambang, telling her that she would receive two sacred items (*gaukang*) from him that would serve as vessels for his spirit and that of his wife. These items soon appeared in the form of golden birds that flew through the air and dropped into her lap. They have been the focus of the royal ancestor cult of Tanaberu ever since (see Gibson 2005: 198–226).

Hasanang was also a niece of a woman from Ara who was named Daeng Gaukang after the golden birds, and who was trained to dance in their honor. Daeng Gaukang was an old woman in 1988 and refused to speak to me, as she feared all *Belanda*, Hollanders. The keeper of the golden birds in the 1940s was Daeng Mangelo, a second cousin of Daeng Gaukang, who also lived in Ara. By arranging this marriage for his son in Tanaberu, Gama thus acquired a whole new set of noble in-laws in Ara as well. Not long afterward, Gama took a second wife for himself from this family when he married the sister of Palolang, Daeng Mangelo's son-in-law.

Between 1947 and 1949 the colonial government sent about 3,000 pilgrims a year to Mecca from the State of East Indonesia as part of their effort to win over the hearts and minds of the population during the counterinsurgency campaign.

> The Netherlands Indies government had special reason to speculate upon this as the government of the Republik Indonesia in Jogja, despite ambitious plans, was not able to allow pilgrims to leave her territory for Mecca; this was caused by the Dutch navy's blockade of the Indonesian ports and by the fact that the RI had no ships available to transport the hadjis. (Vredenbregt 1962: 109–110)

In 1948, Haji Daeng Parani was the first person in Ara to be selected by the government to perform the *hajj*. He was the son of Kali Baso Daeng Siahing and had been serving as Gama's *sulewatang*. No one had made the *hajj* from Ara since Haji Daeng Mareha had accompanied Haji Baso Daeng Raja from Bira in 1895. Gama was selected by the NIT government to perform the *hajj* in 1949. He was accompanied by the *karaeng* of Lemo Lemo, Masalolang, and the *karaeng* of Kajang, Bapa Daeng Matasa (1928–1949) both of whom had also shown their loyalty to the NIT government.

When Gama left on the *hajj*, his eldest son, Padulungi, was installed as *karaeng* of Ara. According to Daeng Pasau, only the nine electors designated in 1938 were allowed to vote, and they chose Padulungi as *karaeng* over two rivals, Patiroi and Daeng Pasau himself, who was only twenty-two years old at the time. Padulungi was formally installed as *karaeng* in the "traditional" manner by the *katte* of Ara, at that time Daeng Mappaccing. He was placed on the *possi' tana*, the stone marking the village's "navel of the earth," and was made to swear an oath while the men of the village brandished weapons around him. The grandson of the mysterious stranger from Selayar, Panre Abeng, had successfully asserted his right to be installed as the rightful heir of the local royal ancestors.

Conclusion: Cosmopolitan Piety and the Late Colonial State

The Government of the Netherlands East Indies gradually consolidated itself in the first half of the nineteenth century on the model of British colonialism in India. As the power of local Islamic rulers gradually ebbed in the nineteenth century, so too did the charismatic religious authority that had been associated with the person of the king during the seventeenth century, and with court-sponsored *ulama* and *shaikhs* who had been trained in the Hejaz during the eighteenth century. The way was paved for the acceptance of the colonial state as a purely functional institution without religious significance, and for the acceptance of religion as one of the few spheres of native life outside direct colonial control.

Between about 1860 and 1930, the Dutch colonial state saw itself as intervening in the native societies of the Indies so as to foster one or another sort of "progress." In the earlier, "liberal" phase that lasted until 1900, the objective was primarily to produce the conditions for the development of a capitalist economy, complete with free markets in land and labor. In the second, "ethical," phase, ideas about how to foster native moral and social progress came to the fore. On the whole, however, the notion of progress remained just that, a notion. "In the final analysis, the Ethical Policy did not lead to improved native welfare, and there was a real decline after 1930" (Ricklefs 1981: 148).

The next period was marked by an abrupt turn toward conservatism, one in which the colonial government largely abandoned any notion of rapid progress among the native population. In the 1930s, someone in Gama's position who had no feudal pedigree and who had established his hegemony by attacking feudal rituals, faced the prospect of being turned out of power. He responded by appropriating the cult of the village *shaikh*, Bakka' Tera', for his own family, and by shifting his allegiance from the party of Islamic reformism, now identified with the dangerous nationalist Nape Daeng Mati'no, to the more traditionalist *karaeng* of Bira, Andi' Mulia. He then uncoupled the cult of Bakka' Tera', which he could control, from that of Karaeng

Mamampang, which he continued to repress. The rituals in honor of Bakka' Tera' are remarkably similar to those performed in honor of Karaeng Mamampang by the previous *gallarrang*, Daeng Makkilo. But while only a direct descendent of Karaeng Mamampang could participate in the latter rituals, Gama could approach Bakka' Tera' as a Muslim devotee. By promoting one cult and undercutting the other, Gama was able to reshuffle the local social hierarchy in a fundamental way.

In summary, we can see that by 1950, Abeng's descendents had gradually consolidated a remarkable degree of control over the local religious, political, and social orders. First, Abeng asserted his superiority within the purely religious order by calling into question the Islamic piety of the local rulers. For example, he insisted that when he died he should be buried in a new, purified graveyard. This wish was fulfilled in about 1910. Second, Abeng's son, Gama, integrated himself into the political order by becoming the client of the Dutch appointed rulers of Bira, who appointed him as the chief of Ara in 1915. When those rulers were removed from office in 1920, Gama became a client of the Dutch *controleur* himself. Third, during the 1930s Gama manipulated the local social order by undermining the cult of the royal ancestors that underpinned the local nobility that was resisting him. In its place, Gama promoted the cult of the Ara's founding Islamic *shaikh*, Bakka' Tera'. Gama secured higher rank for his family in the social order by marrying his son Patoppoi into a noble family in Tanaberu in the 1940s. This family was in control of a rival ancestor cult focused on ritual objects associated with To Kambang. Although Gama himself never approved of this cult, his descendents have become enthusiastic participants. By marrying into this line, Abeng's grandson Patoppoi finally consolidated the family's position within the political, religious, ritual, and social orders.

Abeng and Gama thus adopted a number of different identities during their lives, each of which was situationally useful but each of which also redefined the goal for which their house was striving. Their identities remained unstable even after their deaths, as the divergent trajectories of their descendents required them to rewrite the family history in a number of different ways.

*   *   *

I opened the chapter with an anecdote from my field notes about Haji Basri's attempt to redefine Panre Abeng as a nobleman of province-wide rank. Up until my visit in 2000, all of my respondents had agreed that Panre Abeng was a humble goldsmith from Selayar who probably spent much of his time filling teeth. Such origins were increasingly out of line with the eminence attained by his decedents. Panre Abeng's son, Haji Gama, ruled the village of Ara with an iron rod from 1915 to 1949. Haji Gama's eldest son, Padulungi, ruled the village from 1949 until 1961. His second son, Andi' Palioi, acquired a higher education in the 1950s, married a Bugis noblewoman from Suppa, and was appointed regent of Selayar. His third son, Patoppoi, had married into a noble family in Tanaberu. Patoppoi's son Andi' Azikin became the *camat* of the islands of Jampea and Bonerate before being appointed one of four assistant governors for the province of South Sulawesi.

Haji Basri was the son of Padulungi. He had attended the Teacher's Training College (IKIP) in Ujung Pandang and obtained a postgraduate *doctorandus* degree in history and anthropology. He served as one of Usman Pelly's main informants during his research on boat building in Ara during the 1970s, and collaborated with Professor Abu Hamid of the Universitas Hasanuddin when the latter worked in the acquisitions department of the museum in the old Fort Rotterdam in Ujung Pandang. When I first met him in 1988, he was the head of educational planning for the regency of Bulukumba. By 2000, the family's frame of reference had expanded from the village of Ara to include the whole of South Sulawesi. It seemed only proper that their common ancestor should have come from a noble family with a province-wide reputation. By linking his great grandfather to the high nobility of Luwu' rather than of Selayar, Haji Basri was staking a claim to membership in a province-wide status hierarchy. His attempt to rename him "Tenri" Abeng may have been inspired by the fact that a man from Selayar called Tenri Abeng had been appointed "Minister without Portfolio of Efficiency of State Companies" by Interim President Habibie in 1998 (van Dijk 2001: 546). By naming Abeng after a cabinet minister, Haji Basri was also linking himself to the top of the national political hierarchy. Finally, his citation of an archival document in his revised epitaph for Abeng drew attention to his acquisition of a *doctorandus* degree from the leading educational institution in eastern Indonesia.

Haji Basri's view of Panre Abeng may be contrasted with Muhammad Idris's view of the same man in 2000. Idris was the grandson of Panre Abeng by his daughter Sadaria. As we will see in the chapter 7, he fought on behalf of the Darul Islam guerilla movement from 1952 to 1961 and was never seriously wounded. In 2000, he attributed his miraculous survival to a book of magical spells that had once belonged to his grandfather, Panre Abeng. When Panre Abeng died in 1910, the book had passed to his eldest son, Haji Gama. When Haji Gama left on the *hajj* in 1949, he gave it to his eldest son, Padulungi. It fell into the hands of Haji Gama's nephew, Muhammad Idris, when Padulungi fled to the city in 1954 to escape the insurrection.

Muhammad Idris believed that the spells in the book were so powerful that they had made him virtually invulnerable throughout the rebellion. His clothes had once been stripped off his body by a machine gun, but his skin was left unmarked. Musing on the incident with me in 2000, Muhammad Idris reasoned that the power of the book was so great that it must have originally belonged to the founding *shaikh* of Bira, Haji Ahmad the Bugis, locally known as *Panre Lohe*, The Great Sage (see chapter 3). Since mystical knowledge of this sort can only be safely used by those of the correct lineage, Muhammad Idris reasoned further that Panre Abeng must have been descended from Haji Ahmad, and that he was, therefore, also a member of this lineage himself. He was intensely interested in the *silsila* of Haji Ahmad the Bugis, which I finally obtained from Bira in 2000. He lost interest as soon as he realized that it recorded master–pupil relationships as opposed to father–son relationships. As a good Islamic modernist, Idris rejected the idea that esoteric knowledge could be obtained from a human master. But he did believe in the genealogical transmission of charisma, and in the power of sacred texts.

Idris advanced a number of other arguments that Panre Abeng must have belonged to Haji Ahmad's lineage, despite the absence of a genealogy to prove it. He pointed out that Abeng had given all of his children pious names: Adam, the first Prophet; *Gama*, "Religion"; and *Sadaria*, "Penitence." Adam then named his own son *Baso Iman*, "Faith," and his daughter *Be'ja*, "Wisdom." Now, there was a *kali* of Bira in the mid-nineteenth century named Makota Daeng Pulana *Tu ri Takabere*, "He who Leads the Prayer Service" (see figure 3.2). Daeng Pulana's brother was also called Baso Iman, and the latter's wife was called Be'ja. Idris reasoned that Adam must have named his children after the latter couple, and so there must have been some kind of family relationship between his father Abeng and the *kali* of Bira. While he accepted that Abeng was not the son of Baso Iman, he thought it more than likely that he was his nephew. The best candidate from the Bira genealogies for Abeng's father is another brother of Kali Makota Daeng Pulana called Daeng Manganja. By this somewhat strained logic, then, Idris could portray himself as a direct descendent of Shaikh Ahmad the Bugis. Where Haji Basri sought to link himself to a national political hierarchy through Panre Abeng, Muhammad Idris sought to link himself to a global religious hierarchy.

Chapter 7

Revolutionary Islam and the Nation-State, 1900–1965

Between 1952 and 1961, many of my acquaintances in the village of Ara were involved in the Darul Islam movement, a movement that wanted the Republic of Indonesia to be founded on Islamic principles. The origins of the form of Islamic knowledge advocated by this movement can be traced to a decision taken by the colonial government in the early twentieth century to educate a cadre of Indonesians to staff the lower ranks of the expanding state bureaucracy. The introduction of modern schools and of mass media such as printed books, journals, and newspapers introduced the newly educated Indonesians to the modernist ideas that had been developed in the late nineteenth century by Muslim intellectuals in colonial India and Egypt. The government found it harder to confine the reformist impulse of this generation to purely religious matters than the generation born during the late nineteenth century. The educated Indonesians who were drawn to Islamic modernism played a key role in the birth of Indonesian nationalism during the 1910s and 1920s. As we saw in chapter 6, the government responded to Islamic nationalism by adopting the explicitly reactionary policy of trying to revive the traditional authority of the royal houses they had just spent a century trying to undermine.

Dutch attempts to restore colonial rule after World War II were fiercely resisted in South Sulawesi, which officially joined the independent nation of Indonesia in August 1950. When nationalist guerilla forces led by Kahar Muzakkar were denied the place they expected in the new national army, and when the new state was established on a secular rather than on an Islamic basis, the guerillas withdrew into the bush and declared their support for a modernist Islamic state. Many of my informants were involved in this guerilla force, which was known as the Darul Islam-Tentara Islam Indonesia (DI-TII). In the villages, the insurrection sometimes took on the character of a civil war between the advocates of strict *shariah* law and the adherents of certain ancestral traditions. A neighbor of mine in Ara, Muhammad Idris, spent ten years in Kahar Muzakkar's personal bodyguard. He told me that he had been inspired by the examples of Shaikh Yusuf and Datu Museng, and made resolute by his faith in the esoteric knowledge that made him invulnerable to penetration by blades or bullets. But Idris was also a schoolteacher who combined these charismatic models with the documentary approach to knowledge of the bureaucratic state.

Documentary Knowledge and Disciplinary Technique

Beginning in 1905, the introduction of formal schooling and print technology transformed religious and political life in South Sulawesi. Schools taught people how to produce the explicit sort of information that can be communicated to an anonymous audience through mass media such as newspapers, textbooks, mass meetings, radio, and television. The prototype for this form of knowledge is the document produced by one bureaucrat for another; hence I refer to it as "documentary knowledge." The Dutch had been producing this kind of knowledge in Indonesia since the early years of the VOC. It only became available to most Makassar after the introduction of village schools in the early twentieth century.

Schools were introduced into South Sulawesi by the colonial state in order to expand the numbers of bureaucrats able to produce and consume the vast amount of documentary information needed to run a modern economy and society. The methods used in these schools grew out of early nineteenth-century experiments in mass education in Europe and in some of the British colonies. A new kind of utilitarian Christianity was central to these early experiments, one that eschewed the niceties of doctrinal debates over purely theological questions and focused instead on how moral education could be used to instill sobriety and self-discipline into the laboring classes.

Michel Foucault linked documentary knowledge and power to the development of a whole host of "disciplinary" methods that were developed to train the bodies and souls of individuals in armies, schools, and factories.

These small techniques of notation, of registration, of constituting files, of arranging facts in columns and tables that are so familiar to us now, were of decisive importance in the epistemological "thaw" of the sciences of the individual. . . . [One] should look into these procedures of writing and registration, one should look into the mechanisms of examination, into the formation of the mechanisms of discipline, and of a new type of power over bodies. (Foucault 1977: 190–191)

Timothy Mitchell has noted that many of the disciplinary techniques mentioned by Foucault were first developed in a colonial setting, as a handful of Europeans struggled to administer vast overseas empires with the help of local subalterns. More specifically, he has applied Foucault's analysis to colonial Egypt, showing how traditional systems for transmitting sacred Islamic knowledge were replaced by radically new systems when schools were introduced (Mitchell 1988: 35). Like many of Foucault's followers, Mitchell tends to emphasize the way modern disciplinary techniques increase the power of the ruling class. But he also notes that their introduction often has unintended consequences: "The schools, universities and the press, moreover, like the military barracks, were always liable to become centers of some kind of revolt, turning the colonizers' methods of instruction and discipline against them" (Mitchell 1988: 171).

In fact, the introduction of schools into colonies all over the world almost always led to the emergence of an anticolonial movement. This is reminiscent of Marx's argument that the Industrial Revolution would strip inherited skills from the bodies and souls of particular workers and embed them in new sorts of machinery. As workers became alienated from particular labor processes, they would come to see themselves as interchangeable members of a universal class. Something similar happened in colonial schools. As students were trained in the production and consumption of generic forms of documentary knowledge, the cultural and linguistic boundaries between them broke down and a national consciousness arose.

This analogy is more apt than it may seem at first, since early efforts at mass education were inspired by the new system of factory production. The monitorial method was first hit upon in 1789 by Andrew Bell. If we take a careful look at the social background of men such as Bell and other early advocates of his "monitorial method," we will see that it was not simply the expression of a repressive policy that aimed at maximizing European domination of the colonies. It was also the expression of a revolutionary struggle against feudal domination in Europe, and this method quickly became a weapon in the struggle for national liberation in the colonies.

Bell was the son of a barber who was famous for his mechanical skill. He studied at the University of Saint Andrews, spent some years in Virginia, and considered a career as a Presbyterian minister. He converted, however, to the Church of England and was ordained in 1784. He sailed for Madras in 1789 to take up a position as headmaster of a school for the orphan boys of soldiers. After recording Bell's frustration with the indolent British teachers at the school, his biographer tells the following story.

> Things were in this state, when, happening on one of his morning rides to pass by a Malabar school, he observed children seated on the ground, and writing with their fingers in the sand, which had for that purpose been strewn before them. He hastened home, repeating to himself as he went "Eureka, I have discovered it," and gave immediate orders to the ushers of the lowest classes to teach the alphabet in the same manner, with this difference only from the Malabar mode, that the sand was strewn upon a board. (Southey and Southey 1844: 172)

Commenting on this passage, Salmon notes that it was the resistance of the British teachers "to try a device picked up from the natives" that forced Bell to use the older boys to teach the younger ones (Salmon 1932: xviii).

Bell's methods were picked up by Joseph Lancaster, who read Bell's work in 1800 and went on to found the monitorial movement. Like Bell, Lancaster was the son of a manual laborer, in his case, of a sieve maker. Like Bell he was originally interested in a career as a dissenting minister. Instead of turning toward the high church like Bell, however, he joined the Quakers. Lancaster publicized the monitorial method, which soon attracted adherents all over the world.

The whole notion of popular education promoted by these working-class pedagogues was fiercely resisted at the time by political conservatives.

> The strong English reaction against the French Revolution made all social innovation suspect. An important article of this conservative faith was the danger of educating the poor. Education would make them unhappy with their lowly station; it would enable them to read seditious literature; it was unnecessary and unwise in a properly hierarchical society. Writing and arithmetic were thought even more inappropriate than reading. This was the backdrop against which new advocates of mass education had to struggle as England entered the nineteenth century. (Kaestle 1973: 2)

As we have seen, men such as Bell and Lancaster came from decidedly humble backgrounds and saw the extension of education in the colonies in the same reforming light as the education of the working classes in England. The monitorial method of schooling was thus originally introduced in England by political radicals from working-class backgrounds who wished to extend access to knowledge to the entire population. The new methods of mass education posed a threat not only to the upper classes in Europe but to existing forms of religious knowledge in Asia. Another early enthusiast of the Lancaster's system was Stamford Raffles. Raffles was born at sea in 1781 to a merchant captain who never prospered. He had to leave school at the age of fourteen and had to begin work as a clerk in with the East India Company. He worked his way up the ranks, culminating with his service as Lieutenant Governor of Java from 1811 to 1816. After Java was handed back to the Dutch, Raffles was appointed lieutenant governor of the colony of Bencoolen on the west coast of Sumatra. Soon after his arrival, he charged a committee headed by the colony's chaplain to consider how Lancaster's system might be introduced to educate the natives (Wurtzburg 1954: 548).

Peter van der Veer has shown that the extension of schooling in Britain's Asian colonies was originally associated with the rise of evangelical nationalism in Britain during the first half of the nineteenth century, when British identity became closely tied to the notions of moral progress and the duty to bring Christianity to the benighted heathens in the empire. This religious definition of the nation was only gradually replaced by a racial definition in the second half of the nineteenth century, in part because the "heathen" proved so recalcitrant to conversion, at least in Asia (van der Veer 1999, 2001).

Schooling and Islamic Modernism

During the nineteenth century, many of the same pedagogical techniques used in European schools were adopted for the transmission of Islamic knowledge, especially where the imposition of direct colonial rule had undermined the role of local royal courts in the transmission of Islamic learning. These techniques transformed the nature of Islamic knowledge and led to a new kind of critique of traditional ritual techniques for communicating with spirits of local ancestors and Islamic *shaikhs*, and with other invisible beings such as *jinn*. Muslims who had been through modern school systems began to read the scriptures as if they were bureaucratic documents such as legal codes, textbooks, and newspapers that contained clear a precise list of rules and regulations for the conduct of life and unambiguous information about God and his creation.

The Muslims of northern India were among the first to experience the loss of political power to Europeans and the need to create religious institutions that were independent of the state (Robinson 1993). One of the first institutions to implement a standardized curriculum and graded classes was the *madrasa* at Deoband that was established by Rashid Ahmad Gangohi (1829–1905) and Muhammad Qasim Nanautawi (1833–1877) (Metcalf 1982). At about the same time, a charismatic Persian, Jamal al-Din "al-Afghani" (1839–1897), and his younger Egyptian disciple, Muhammad Abduh (1849–1905), were inspired by Enlightenment critiques of Christianity as contrary to reason. They used similar arguments to revive the rationalistic views held by the Mu'tazila and Faylasuf in the first centuries of Islam, and to argue that Islam was the only religion in which there was no conflict between reason and faith.

Abduh wrote in support of the nascent Egyptian nationalist movement throughout the 1870s. When the British seized control of Egypt from the Ottomans in 1882, Abduh was exiled to Lebanon. There he experienced a spiritual crisis that turned him in the direction of religious reform rather than political activism. He was allowed to return to Egypt in 1888 and developed a cordial relationship with the British consul-general, the Earl of Cromer (r. 1883–1907). He was appointed as a judge in one of the new courts of positive law. In 1899, he was appointed *mufti*, or chief jurisprudent, of Egypt. He used his influence with the colonial administration and his status within the traditional religious system to lead a reform of the Islamic educational system from village-level schools up through the university of al-Azhar (Hourani 1962).

Gregory Starrett has presented a nuanced argument concerning the intended as compared to the actual effects of the introduction of European schooling into Egypt (Starrett 1998). The intended purpose of mass education both in England and in the colonies was the production of a disciplined and obedient workforce. The Bible was used in English schools only as a source of edifying examples from which practical moral lessons could be derived. In Egypt, the government tried to adapt the existing system of Koranic schools by reducing the scriptures to an "objectified" set of doctrines and a "functionalized" list of moral rules.

> In essence, the functionalization of Egypt's religious tradition meant that the ideas, symbols and behaviors constituting "true" Islam came to be judged not by their adherence to contemporary popular or high traditions, but by their utility in performing social work, either in furthering programs of social reform or in fulfilling the police functions that Europeans attributed to education as such. (Starrett 1998: 62)

These schools soon created a new Egyptian elite. Rather than serving as dependable agents of the colonial government, however, this elite led the resistance to the British occupation of Egypt. By the 1930s, schoolteachers such as Hassan al-Banna had built a modern political organization based on institutions such as schools, youth groups, newspapers, and national congresses, all dedicated to the cause of religious nationalism. Al-Banna's organization, the Muslim Brothers, took over the British charge that existing Islamic institutions were "dry, dead, ritualistic, and irrelevant." But they accused the British of having allied themselves with the obscurantist Islamic establishment based in al-Azhar. The British replied by accusing the Muslim Brothers of being obscurantists allied with the royal palace. Different groups could thus appropriate an "objectified" and "functionalized" Islam for their own political ends (Starrett 1998).

Egyptian modernism found a ready audience in Indonesia among those who had been exposed to the new educational methods. Modern schooling was introduced on a limited basis throughout Indonesia in the early years of the twentieth century. Governor-Generals van Heutzs (1904–1909) and Idenberg (1909–1916) favored practical education in the vernacular. In 1907 van Heutzs set up village *volkscholen* with a three-year course of study. In 1921 these were linked into a system of schools, although these remained inaccessible to the poor. The depression of 1930 brought these early attempts at mass education to a halt. Literacy at that point was still only 7.4 percent throughout the colony, and much of that was attributable to Koranic schools. The highest literacy rate was 50 percent in South Maluku, due to the presence of Christian mission schools. By comparison, 25 percent of the population of the Philippines had been taught to speak English in American schools by 1939.

Academic advisers to the government such as Snouck Hurgronje and Abendanon favored educating an elite group of Indonesians in Dutch so that they could take over civil service positions, balance Islamic fanaticism, and ultimately inspire the lower levels of society to imitate European ways. First-class schools for the elite were reformed in 1907 and consisted of a five-year course of study. In 1914, graduates were allowed to attend Dutch High Schools for the first time, and in 1919 high school graduates were allowed to go on to study at Dutch Universities.

A parallel network of modern Islamic schools was set up around the same time that the government school system was being organized. The most important modernist Muslim organization in Indonesian was the Muhammadiyah, founded by Ahmad Dahlan (1868–1923) in

Yogyakarta, Java in 1912. Ahmad Dahlan went to Mecca in 1890, where he studied under the famous Minangkabau scholar Ahmad Khatib (1860–1915). It was also in Mecca that he was exposed to the ideas of Muhammad Abduh.

The Muhammadiyah grew slowly at first. By 1925, it had recruited only 4,000 members. The subsequent growth of the organization throughout Indonesia was due in large parts to the efforts of a Minangkabau, Haji Abdul Karim Amrullah, better known as Haji Rasul (1879–1945). Like Ahmad Dahlan, Haji Rasul studied in Mecca under Ahmad Khatib, from 1901 to 1906. Upon his return to Sumatra, he established an Islamic school in Padang Panjang "along modern lines with a system of graded classes with desks" (Laffan 2003: 171). The graduates of this school became known as the *Kaum Muda*, the "young upstarts" in Laffan's rendering (Laffan 2003: 234). A formal union between the Sumatran and Javanese modernists took place after Ahmad Dahlan's death in 1923, and the Muhammadiyah rapidly developed into an Indonesia-wide organization thereafter. By 1938, it had 250,000 members, 1,774 schools, 834 mosques, and 31 libraries located all over Indonesia.

The pioneer of Islamic modernism in South Sulawesi was Haji Abdullah bin Abdurrahman of Marusu'. Like Ahmad Dahlan and Haji Rasul, he spent several years studying in Mecca, in his case from 1907 to 1917. Upon his return he founded an Islamic school in Makassar "in which the teaching was decidedly reformist." In 1923, he founded an organization called As-Sirath al-Mustaqim. In 1926, this group merged with the Muhammadiyah. The first branch of the Muhammadiyah outside the city of Makassar was formed in Wajo' in 1928 (Pelras 1985: 127).

The Persatuan Islam was founded in the early 1920s in Bandung, west Java. Persis was devoted to the propagation of modernist ideas "by holding public meetings, *tabligh*, by conducting sermons, study groups, organizing schools and publishing pamphlets, periodical and books" (Noer 1973: 85). Its publications served as references for Muhammadiyah members. It published the journal *Pembela Islam* from 1929 to 1933, and *Al-Lisan* from 1935 until the Japanese occupation. They were widely read in Sulawesi (Noer 1973: 91).

One of the earliest organizations to link modernist Islam with nationalism in Indonesia was the Sarekat Islam (SI), which was founded in 1911 by a group of merchants in Solo, central Java. In its early years, nationalists, socialists, and Islamic modernists all coexisted within the SI. In 1921, the SI expelled all members of the Communist Party of Indonesia (PKI) and its membership declined (Ricklefs 1981: 166).

In 1929 the SI purged all members of the Muhammadiyah for maintaining too close a relationship with the colonial government, and renamed itself the Partai Sarekat Islam (PSI) (Ricklefs 1981: 168). The PSI continued to decline in membership and influence during the 1930s, as did the Indonesian Nationalist Party founded by Sukarno in 1927.

Ataturk's efforts to secularize the Turkish state provoked an open split between the followers of the Egyptian-inspired modernism of Dahlan and those who held to more traditional practices.

> In 1924 Turkey abolished the position of Caliph, the spiritual head of all Muslims, which the Ottoman Sultans had claimed to be for some six decades. Egypt planned an international Islamic conference to discuss the caliphate question. But further confusion ensued when in 1924 Ibn Saud captured Mecca, bringing with him puritanical Wahhabi ideas of reform and a claim that he was Caliph. He, too, invited all Muslims to a caliphate conference. During 1924–6 Indonesian Muslims set up committees to attend these conferences but the representatives were predominantly Modernist, and Tjokroaminoto appeared prominently. (Ricklefs 1981: 168)

The traditional Shafii *ulama* of Java regarded both the modernist Muhammadiyah and fundamentalist Wahhabi positions on many matters as little short of heretical. Rejecting the modernist claim to represent all Indonesian Muslims they set up the Nahdlatul Ulama (NU) to represent their traditionalist point of view. By 1942 the NU had 120 branches, most of which were located in Java (Ricklefs 1981).

The Muhammadiyah in South Sulawesi

In 1931, the Muhammadiyah held its Twenty-First Congress in Makassar. Haji Rasul sent his brilliant young son, Haji Abdul Malik Karim Amrullah (Hamka) to Makassar to prepare for it. Although he was only twenty-three years old, Hamka was enthusiastically welcomed as a great teacher by local members of the Muhammadiyah, who begged him to stay on in Makassar when the Congress was over. He remained until the end of 1933, making numerous trips around the province to spread the modernist message. In later years he often used Bugis and Makassar customs as examples of pagan practices that had to be fought, including drinking, gambling, and cockfighting, the payment of high bride prices, and the veneration of sacred places.

During his two years in Makassar he published two journals and a book. Hamka introduced a new, popular style to persuade his audience of the need to abandon bad customs (Steenbrink 1991: 228). By the time Hamka left Makassar, there were sixteen branches of the Muhammadiyah in South Sulawesi. In 1937, there were sixty-six branches (Alfian 1969: 465). By the time of the Japanese invasion in 1941, the Muhammadiyah could boast 7,000 members and 30,000 sympathizers in South Sulawesi (Pelras 1985: 127).

The Attack on the Cult of the Prophet

The Muhammadiyah launched a number of campaigns against what it regarded as idolatrous practices in South Sulawesi, beginning with the recitation of the *Barasanji*. As we saw in chapter 5, the *Barasanji* became a key component of village life during the nineteenth century. What modernists found most objectionable in the recitation of the *Barasanji* was that participants rose to their feet at certain points to show respect for the spirit of the Prophet, a practice known in Malay as *berdiri maulid*. The controversy illustrates the extent to which a literate public had developed in Malaysia and Indonesia, tied together by the publication of brochures, booklets, and newspaper articles in which debates developed with great rapidity. The debate exposed a large audience to the modernist methodology of rejecting all argument from tradition and relying solely on the Koran, on strong *hadith*, and on individual reason. Since modernists refused to attend the life-cycle rituals during which the *Barasanji* was recited, the controversy disrupted social life all over Indonesia.

The controversy reached Southeast Asia in 1906, when the Egyptian reformist journal *Al-Manar* published a letter from an inhabitant of Johore concerning the *Maulid* of al-Dabi' and the practice of standing during its recitation. In reply, not only was this practice attacked, but the value of reciting the text at all was called into question. This provoked a furious response from Singapore, which was published by *Al-Manar* in April 1906. The journal stood by its position, however, of condemning the practice outright. In November 1906, *Al-Imam*, a newly established reformist journal in Singapore, also declared the *berdiri Maulid* an illegitimate innovation. In 1909, Haji Rasul published a lengthy poem in which he condemned the practice of standing in honor of the Prophet, but not the recitation of the text itself.

In 1919 the controversy reached a climax when a series of public debates were held in Sumatra. The modernists were represented by Haji Abdullah Ahmad, who had been the publisher of the journal *Al-Munir* (1911–1916) and the traditionalists were represented by Shaikh Chatib Ali. Significantly, B.J.O. Schrieke of the Bureau for Native Affairs presided over the second debate. Shaikh Ahmad Chatib later published a letter from Schrieke in which he stated that he agreed with the traditionalist position. But in 1921 Schrieke denounced a brochure reissued by the traditionalists as "testifying to an even greater degree of falsehood" than the original edition, and approved the version of the debate published by the reformer Haji Abdullah Ahmad (Kaptein 1992).

Although the outcome of the debate was inconclusive, the very fact that religious scholars had submitted their views to the judgment of the general public established an important new principle: nothing should be done simply because it had always been done. All traditional ritual practices should be exposed to explicit, public debate and justified through the rational examination of textual evidence.

The split between the modernist and traditionalist positions on the *berdiri maulid* was formalized in 1930 when the Fifth Congress of the NU declared that the *berdiri Maulid* was a "legally accepted custom which was recommended." Two years later, the modernist condemnation of the practice was promulgated in Makassar during the twenty-first Congress of the Muhammadiyah.

The Attack on the Cult of the Royal Ancestors

The Muhammadiyah also launched a concerted attack on the cult of the royal ancestors in South Sulawesi (Chabot 1950: 86–87). These cults were centered on the *gaukang*, powerful objects that played a crucial role in the installation and legitimacy of new rulers. Every ruling family had to make regular blood sacrifices in honor of the *gaukang*. As schooling and Islamic modernism spread, many nobles became reluctant to continue these practices. Ironically, this occurred at the very moment when the Dutch administration was starting to promote hereditary kingship and royal ritual again as a hedge against nationalism.

The campaign against the cult of the royal ancestors reached the Konjo Makassar when the ruler of Kajang died in 1928. He was to be succeeded by his son, Karaeng Yahya Daeng Magassing. Karaeng Yahya had come under the influence of the Muhammadiyah, however, and refused to apply sacrificial blood to the *gaukang* of Kajang. Instead, he reached into the sacred bundle, pulled out the head of a walking stick, and displayed it derisively to the assembled elders. They were deeply shocked and reported him at once to the authorities. Karaeng Yahya was deposed and Bapa Daeng Matasa was appointed *karaeng* of Kajang in his place. Hamka later became friendly with Karaeng Yahya and praised his actions in his commentary on the Koran (Hamka 1965–1982 IX: 248–249).

Schooling and Islamic Nationalism in Ara and Bira

Until 1920, the only schools in Bulukumba were three-year *volksscholen* in Bulukumba and Kajang. In 1922 the Dutch opened *volksscholen* in Bira and Kalumpang, and, in 1925, in Ara. Ramalan Daeng Pabuka was the first teacher in the village. He had been trained in Kajang. Instruction was in Malay. The children were first taught the *lontara* script, and then from the second grade, the Roman alphabet. By 1941 there were still only seven six-year schools with eight hundred and sixty-five pupils in all of Bonthain Division (Harvey 1974: 92). Until 1955, children from Ara still had to go to the *vervolgscholen*, "continuation schools," in Bulukumba City, Kajang, or Tanete to complete a basic five-year education. One of the first students from Ara to attend a government school was Ebu. He started school in Bulukumba, and then took teacher training course in Segiri, Pangkajene. He returned to teach in Ujung Lohe in 1944.

Beginning in the 1920s, there was also an Islamic school run by the Muhammadiyah in Bulukumba City. Among those who attended were Pantang Daeng Malaja of Ara (1894–1984); Andi' Abdul Karim Daeng Mamangka (r. 1938–?) and Andi' Lolo Tonang of Tiro; and Nape Daeng Mati'no of Bira (ca. 1890–1954; r. 1942–1950). It was Daeng Mati'no who was responsible for introducing Islamic nationalist ideas to the villages of the Bira peninsula. He absorbed these ideas from the Sarekat Islam during the period from 1921 to 1931 when the colonial government exiled him to Java. During the 1930s, his loyalties appear to have shifted to the Persatuan Islam. Collins mentions that the second son of Uda Daeng Patunru, probably Nape Daeng Mati'no, went to Bulukumba to attend a meeting of the "Parsi" organization in 1935 or 1936 (Collins 1936: 284). He also mentions "Parsi" members as having preached to large crowds in Tanaberu, where the current *kepala district* had the speaker arrested; and in Balangnipa, where the Dutch commandant of the *afdeeling* of Bonthain had the speaker arrested. The first speaker challenged Muhammadiyah positions on certain religious laws, such as the ban on intercourse during the daytime in the month of Ramadan. The second speaker told his audience not to fear weapons, only Allah (Collins 1936: 242–243).

Muhammad Nasir told me that Daeng Mamangka of Tiro had belonged to the PSI in his youth. When he became the *karaeng* of Tiro in 1936, he proceeded to implement many modernist doctrines derived from the Muhammadiyah, which continued to grow throughout the 1930s. Its strategy of avoiding overt political agitation allowed it to provide a framework within which a new generation of religious nationalists could be formed without alarming the authorities.

The ties between Islamic modernism and Indonesian nationalism were only strengthened by the Japanese, who "established a Department of Religious Affairs, largely for Muslim concerns; supported the creation of a unified Muslim political federation, known as Masyumi; and, eventually, as the threat of allied invasion loomed large, trained Muslim militias" (Hefner 2000: 41).

Haji Muhammad Nasir, the assistant registrar of marriages for the new village of Darubia in 2000, told me that Daeng Mati'no had hoisted the Japanese flag in 1942 to welcome the occupation forces. Since he was well known as an Islamic nationalist who had once been exiled by the Dutch, the Japanese appointed him as ruler of Bira, and he installed his brother, Andi' Baso, as *kali*. Ever nimble politically, Gama remained in control of Ara throughout the Japanese occupation.

The War of Independence and the Darul Islam Insurrection, 1945–1961

After Japan's defeat, Indonesian nationalists declared independence on August 17, 1945. The Dutch were intent on reestablishing their colonial authority over Indonesia. Many Bugis and Makassar went to Java to fight the Dutch. Among them was La Domeng, a minor Bugis noble born in 1921 in a village in Luwu'. In 1934 he completed primary school there and in 1937–1940 attended a Muhammadiyah teacher's school in Surakarta, Java, where he took his new name from a favorite teacher, Kahar Muzakkar. In 1941–1943 he returned to Luwu' to teach in a Muhammadiyah school. In 1943 he was banished from Luwu' for denouncing the existing feudal system in South Sulawesi and for advocating the overthrow of the aristocracy. He returned to Java and spent 1943–1945 in business in Solo. From 1945 to 1950 he led a group of guerrillas against the Dutch in Java (Harvey 1974: 181–182, 474). According to Andi' Anthon of Luwu', the fifty-two boats that carried Bugis fighters from Ponro and Jalan in South Sulawesi to join Kahar Muzakkar in Surabaya were built in the boatyards of Luwu' by men from Ara (personal communication).

Beginning in late 1949, the Bugis and Makassar guerrillas who had been fighting the Dutch in Java began to return to South Sulawesi and to form themselves into local battalions. With independence in sight, the question arose of what to do with these irregulars under the new Republic of Indonesia. Many were untrained and uneducated and there was a certain reluctance on the part of the professional officer corps to admit them into the regular army. In June, 1950, Kahar Muzakkar was sent by the new government from Java to Makassar to help resolve this "guerrilla question." Sulawesi was officially incorporated into the new republic on August 17, 1950, five years to the day after Sukarno issued the Indonesian Declaration of Independence.

Kahar Muzakkar fully expected that his men would be inducted into the regular army and that he would be put in command in South Sulawesi. They soon felt, however, that they were being passed over. Kahar Muzakkar withdrew to the *hutan*, the bush, almost as soon as he arrived, and engaged in a long series of inconclusive negotiations with the authorities. A nephew of Haji Gama named Muhammad Idris joined the rebellion in 1951. He served as a member of Kahar Muzakkar's elite commando unit, the *Mobile Moment Comando* or *Momoc*. This acronym was a play on the Indonesian *momok*, bogeyman or ghost, since the guerillas moved invisibly at night like ghosts (Harvey 1974: 406). Muhammad Idris was very reluctant to say very much about the movement in 1988 and 1989 because of the New Order's hostility to any form of political Islam. He was more forthcoming in 2000 after the fall of the regime, and began to frame his account of Kahar Muzakkar in ways that recalled the stories of Shaikh Yusuf and Datu Museng. I will return to his narrative of the movement later.

In August 1953 Kahar Muzakkar declared his support for the Negara Islam Indonesia (Islamic State of Indonesia, NII), His movement became known as the Darul Islam/Tentara Islam Indonesia, Abode of Islam, Islamic Army of Indonesia (DI/TII). By 1953, the guerrillas had gained control of much of the countryside in South Sulawesi, including all of Bulukumba except for the city proper.

The ideology of the Darul Islam movement combined a rigorous interpretation of Islamic law together with extreme hostility to "feudal" practices that set the nobility apart from commoners. Strict *shariah* law was introduced in areas under guerrilla control. Sufi *tariqa*, the veneration of tombs, and the royal ancestor cults were suppressed. As we saw in chapter 5, the *tariqa'* had been closely associated with the royal courts until the late nineteenth century. In chapter 6, we saw that they had spread downward into peripheral villages such as Ara and Bira by the 1930s, but were still used to legitimate the local social hierarchy. The guerillas thus attacked them for both theological and sociological reasons. As part of their campaign against "feudalism," all symbols of differential social rank were suppressed in the performance of life-cycle rituals. Weddings were reduced to the payment of the same minimum sum of *mahar*, dowry, by all grooms and the signing of the *nikah*, the wedding contract, by the two parties before witnesses.

The breakdown of government authority and the absence of an effective DI/TII provisional government fostered the outbreak of family feuds that dated back to Westerling's pacification of the province in 1948. Village officials and schoolteachers fled to the cities for protection. Bulukumba City, which had been a predominantly Bugis enclave in Bulukumba, acquired so many new Konjo inhabitants in these years that the two ethnic groups are roughly in balance to this day.

Independence and Insurrection in Ara, 1950–1954

In 1950, elections with universal adult suffrage were held for the first time in Ara and Bira. Padulungi ran against Baso Daeng Paroto, nephew of the last hereditary Gallarrang, Daeng Makkilo; and against Andi' Paca, a first cousin of Baso Daeng Paroto. The vote for the old ruling line was thus split and Gama's son Padulungi won. In Bira, Daeng Mati'no remained *karaeng* until 1951. He was succeeded by Muhammad Ahmad Karaeng Salle, a nephew of the former regent, Andi' Mulia. Padulungi and Karaeng Sale were not able to rule in peace for long because of the conflict that was brewing between the returning resistance fighters and the new national government.

As the Darul Islam insurrection took hold all over South Sulawesi in 1953, the situation in Ara and Bira degenerated into anarchy. The district chief of Ara, Padulungi, fled to the city in 1953, leaving his twenty-five-year old secretary, Daeng Pasau, in charge. His authority was contested by Daeng Majannang, the nephew of last hereditary *gallarrang*, Daeng Makkilo. Daeng Majannang told me that he had only been able to stand up to the Darul Islam guerillas because of his superior invulnerability magic. Although he had retired from government office when he went on *hajj* in 1949, Haji Gama was still a target of the guerillas as a possible agent of the state. In 1953, they tried to convict him of a technical violation of the *shariah* law so that they could execute him. They accused him of having pronounced the *talak* formula three times on his wife, rendering their divorce final, but of then having sexual intercourse with her. In the eyes of the Darul Islam authorities, this constituted adultery and was punishable by death. As proof, they claimed that his wife had become pregnant after the divorce. The local commander sent his men to arrest Haji Gama, but Daeng Majannang warned him to hide in time.

They then arrested Daeng Majannang for obstruction of justice. They employed a variety of methods to execute him. First they buried him in a hole up to his waist, tied his hands behind his back, and attached them to a horse. They then whipped the horse to wrench him in half, but the horse refused to budge due to a spell Daeng Majannang recited, and afterward he levitated himself gently to the surface of the ground. Next, they tried tying a rock to his leg and throwing him in the water, but he made the rock buoyant. Finally, they tried to stab him with a *keris*, but the blade broke on his skin. It became clear that his invulnerability magic was just too strong for them. At this point, Haji Gama showed up with his ex-wife in the company of the *karaeng* of Tiro. His ex-wife was examined by some women and was proved to be menstruating, so the case was dropped.

This incident proved to be only a temporary reprieve for Haji Gama. He continued to travel freely around the area. One day as he was returning by bicycle from Tanaberu with Daeng Pasau, they were stopped by a unit of the DI/TII. A column of government troops had been spotted marching south from Tiro, and the unit suspected that Haji Gama had been informing on them in Tanaberu. They let Daeng Pasau go but told Haji Gama they were going to execute him. Since it was time for evening prayers, he received their permission to pray. As he bent to the ground, he was shot through the heart, and then his throat was cut.

After his death, Haji Gama's tomb became a source of esoteric knowledge and power. Palippui, a local mystic, told me that while he was meditating on his grave, Haji Gamma had appeared with his head dangling down from where his throat had been cut. Haji Gama told him that the commander who ordered his execution could not kill him with his own gun. To overcome his invulnerability magic, the commander's wife had to shoot him with a golden bullet. The commander was not from the area, but the man who cut his throat was a man he knew from the neighboring village of Kalumpang. The man had taken his *keris* and his ring. Palippui later went to the man's house and spoke to his wife. She admitted that the story was true and that they still had Haji Gama's *keris*.

Like Haji Gama, the former ruler of Bira, Daeng Mati'no, refused to move to Bulukumba City in 1953. Haji Muhammad Nasir of Darubia told me that Darul Islam guerillas from Bonto Tiro executed Daeng Mati'no in 1954 along with his brother, the *kali* of Bira Andi Baso. The village of Tanaberu was burnt to the ground in May 1955, and another wave of government officials fled to Bulukumba. One of my informants, Ebu, was the principal of the elementary school in Ara at the time. He also fled Bulukumba when the Darul Islam fighters began targeting all civil servants for execution. He taught a class of thirty students from Ara in Bulukumba until it was safe to return in 1961.

The Dompe Army of the Amma Lolo, 1954

The Darul Islam's intolerance of local customary practices and the imposition of a strict form of *shariah* law led to a backlash among the devotees of the royal ancestor cults, especially among the followers of the *Amma Towa*, the Old Father, in Kajang. There is a tremendous local mystique around this figure in South Sulawesi. His followers believe he is the reincarnation of all of the previous Amma' Towas, but even pious Muslims suspect him of having great supernatural powers. Many of the Konjo Makassar inhabiting the coastal settlements of Hero, Lange Lange and Kajang continued to regard the Amma Towa as a spiritual leader well into the twentieth century (Usop 1985).

While I was conducting fieldwork in the 1980s, the inhabitants of Tana Towa were officially regarded as Muslim, even if of dubious orthodoxy. A story in a local newspaper in 1989 explained that when Datu Tiro first arrived in the area, he quickly converted the ruler of Tiro, I Launru Daeng Biasa, by performing a number of miracles. The king of Kajang heard of these feats and sent two emissaries, Janggo Toa (Old Beard) and Janggo Tujarre (Faithful Beard) to study Islam with him. Janggo Toa studied with Datu Tiro for only a short time before returning to Kajang. Janggo Tujarre continued his studies with Datu ri Patimang in Luwu' and with Datu ri Bandang in Gowa, but he never acquired a very deep understanding of its key doctrines either. To this day, the inhabitants of Tana Towa hold that in order to live as Muslims one only needs to recite the confession of the faith, and to follow the proper rituals for slaughtering animals, marriage, funerals, and circumcision (Andi' Shadiq Kawu 1989).

In 1954 an adherent of the Amma Towa living in Tanuntung, a village in the realm of Lange Lange, set himself up as the *Amma Lolo*, or the Young Father. He had the support of Karaeng Kilong, the district chief of Lange Lange. The Amma Lolo formed an army distinguished by the traditional head cloth its members wore, folded so that a triangular peak drooped over the top in a style called *dompe*. According to my informants in Ara, the low-lying swampy area that stretches along the river that runs from Tana Towa in Kajang through Hero and Lange Lange to the sea has always been a center for banditry, gambling, and drinking palm wine. It was from this area that the Amma Lolo recruited most of his followers, although some came from as far away as Tanete in the north and Ara in the south. According to Daeng Pasau, the Dompe Army even received some government support at first, since it was so strongly opposed to the Darul Islam movement.

Some time in 1954, the Amma Lolo summoned the leaders of all the coastal Konjo villages to a meeting in Tanutung. Daeng Pasau headed a delegation of sixty men from Ara. A man called Raba headed the group from Tiro. The Amma Lolo made a speech announcing his opposition to the DI/TII, and warning that any of its supporters found in a village under his control would be killed. Daeng Pasau seemed generally sympathetic to the movement, which was, among other things, opposed to the egalitarian tendencies of the Darul Islam guerillas. He denied that the Amma Lolo was against Islam or prayer in mosques, saying that only a few hotheaded supporters had taken such an extreme position. Palippui insisted, however, that the Dompe Army had tried to change the declaration of faith from "I bear witness that there is no God but God and Muhammad is His Prophet" to "I bear witness to the Amma Towa and the Amma Lolo."

The Dompe Army was armed only with swords, spears, and magic. In 1955, it swept down out of the northern Konjo lands all the way to Bira, executing many Darul Islam supporters along the way. The villages split into two factions. The Dompe Army managed to recruit at least three fighters from Ara, including Nacong and Daeng Mangelle, who was one of my informants. My host, Abdul Hakim, was a member of the group that prepared itself to defend the village mosque at Ere Lohe.

News of the chaos the Dompe Army was causing and of their anti-Islamic actions eventually reached Kahar Muzakkar himself. He marched on Kajang at the head of his elite MOMOC unit. Muhammad Idris described to me how they first passed through the old Dutch rubber plantation in Tanete and pacified it to secure its rear. They levied a tax on the inhabitants, but did not harm them. They first encountered the Dompe forces at Panremeo in Bonto Tangnga, and there were many casualties on both sides. The Amma Lolo rallied his forces, but they were decisively defeated in a battle on the slopes of Mount Lembang Gogoso.

According to Muhammad Idris, the Dompe Army was not the Amma Towa's idea, but that once the thing had acquired a certain momentum he felt constrained to give it his blessing. To make sure there would be no further trouble, Kahar Muzakkar took the Amma Towa into custody. Muhammad Idris, who was part of his personal escort, came to know the Amma Towa well. In 1961 the Amma Towa was returned unharmed to Tana Towa, and died soon after. Ironically enough, some say he was killed by government soldiers for having supported the Darul Islam rebellion.

The care Darul Islam sympathizers took to protect the Amma Towa is a bit puzzling, since they showed very little tolerance for those who deviated from their strict interpretation of *shariah* law in other matters. The seemingly pagan Bugis village of Amparita in Sidenreng was similarly spared by the Darul Islam movement. It was surrounded at one point by the Islamic guerillas who threatened to convert its inhabitants by force, but a general who was from the area intervened and prevented this from happening. In 1970, the ruling party of Indonesia, Golkar, recognized these villagers as adhering to an officially recognized religion, a form of Hinduism (Maeda 1984).

Andi' Anthon, a noble from Luwu' who visited Ara in the company of the American anthropologist Shelly Errington, told me that there is a similar village in Luwu', called Cerekang, whose inhabitants have never fully converted to Islam and who are regarded by local people as preserving ancient forms of ancestral knowledge. They, too, were left unmolested by the Darul Islam forces, even though Kahar Muzakkar was born in Luwu' and must have known all about them. I will return to the reasons for this remarkable toleration later.

After the defeat of the Dompe Army, the Darul Islam forces established a relatively stable administration in Ara. Under the Darul Islam regime, the village heads were called by the Islamic term for leader, *imam*, rather than the old noble titles of *karaeng* and *gallarrang*. Ahmad Tiro was appointed the *imam* of the *Desa* of Ara. In 1957, Ahmad Tiro was succeeded as *imam desa* by Bagu and was appointed his Chief of Staff. The *desa* was divided into four *kampong*, each under an *imam kampong*: I in the west under Pananroi, II in the center under Daeng Palawa, III in the north under Leko, and IV in the east, under Cebu. The *imam* were meant to govern according to *shariah* law, and according to three volumes of emergency regulations, copies of which still survive in the village. These men continued to administer village affairs until the restoration of government authority in 1960.

The Darul Islam administration outlawed the veneration of the village *shaikh*, Bakka' Tera', and his wife, Daeng Sikati, on the grounds that it was idolatrous. They pulled down the high wall that encircled their tombs and strictly forbade anyone to visit the simple gravestones that were left in its place. Recitation of the *Barasanji* was also forbidden.

In 1961, Padulungi returned from Bulukumba City and served again briefly as chief of Ara. Later in the year, the fourteen districts of the old subdivision of Bulukumba were abolished and replaced by seven *kecamatan*. The new *camat* of Bonto Bahari designated Haji Gama's old scribe, Daeng Pasau, as the *coordinator kampong* of Ara while the reorganization of the villages was being carried out. In 1962, Ara was split into two *desa*, Ara and Lembanna. For the next five years, Daeng Pasau served as *Kepala Desa* of Ara. Padulungi's daughter, Andi' Anis, was then appointed the head of Lembanna and served in that capacity until 1967.

Many of the educated people who had fought for the Darul Islam movement found a place in the civil service after 1961. Ahmad Tiro, who had been the Imam Desa under the Darul Islam, served as the *Kepala Desa* of Lembanna for a time. Pananroi, who had been the *imam* of *Kampung* I under the Darul Islam became the "Official Recorder of Marriages" (*Pegawai Pencatat Nikah*, PPN) of Ara, equivalent to the old office of *kali*. My host in Ara, Abdul Hakim Daeng Paca, was recruited by the Darul Islam forces in 1957, while he was studying at the teacher training school in Bulukumba. He spent a year in Tanete near Mount Bawakaraeng in command of seventeen men before returning to Ara and marrying his wife, Andi' Sutra Daeng Kebo. Hakim explained that their fathers had promised them to one another in 1947 when he was only nine years old, according to the traditional practice of childhood betrothal followed by a long period before the wedding was solemnized. The wedding ceremony itself was anything but traditional, however. It consisted of only the payment of the minimum amount of *mahar*, Rp 125, and the signing of the *nikah* contract. Hakim explained that it was a general policy of the Darul Islam movement that all marriage ceremonies be cut to the Koranically prescribed minimum in the interest of both religious orthodoxy and social justice. Their eldest daughter, Nurhadi, was born a year later. After his marriage, Hakim stayed on in Ara and taught at the school run by the Darul Islam until the end of the rebellion in 1961. Haji Mustari, the younger brother of Daeng Pasau, taught alongside him.

After the restoration of government authority, Abdul Hakim went to work as a secretary to Daeng Pasau, the village chief, and to Pananroi, the *Pembantu Pencatat Nikah* (PPN), or Assistant Recorder of Marriages, the highest religious office recognized by the government. In 1965 he got a job teaching in the local elementary school, and in 1982 he was appointed headmaster of an elementary school in Bira. There is no doubt in my mind that if his formal education had not been interrupted by the Darul Islam rebellion, he would have acquired a higher degree and risen far within the civil service. As it was, his lack of credentials left him trapped in the village.

Conclusion

Modern methods of education were introduced by the colonial government and by Muslims returning from Egypt and the Hejaz early in the twentieth century. The former hoped to produce a steady supply of lower-level bureaucrats who could help implement the state's modernizing project. The latter hoped to equip Muslims with the intellectual tools they needed to master European technology and to compete with Christian missionaries. The schools ended up delegitimating both the colonial bureaucracy and the secular nationalist bureaucracy that followed it. Unlike Haji Gama's generation of pious Muslims, many men in Abdul Hakim's generation were radicalized by the introduction of modern schooling into the village. It upset many of the received models of what true knowledge was and how it was acquired. Explicit propositional knowledge derived from printed scriptures, textbooks, and newspapers; public debates in which persuasive rhetoric was more important than the descent and official position of the speaker; and point-by-point comparisons between local rituals and cosmopolitan Islamic practice all served to undermine the traditional authority of rulers and religious officials.

This generation would not accept the colonial government in bureaucratic terms as legitimated by its own efficiency, nor would they accept its alliance with a reactionary noble class legitimated by custom and tradition. Instead they embraced the neo-Mu'tazilite arguments of modernists such as Muhammad Abduh of Egypt, who argued that there was no incompatibility between reason and revelation, between science and faith. They adopted a radical approach to all local social and religious practices that could not be justified in the explicit forms of reasoning and argumentation taught in the *madrasas*.

Toward the end of my stay in 1989, I asked Abdul Hakim why he thought that the Darul Islam movement and the Dompe Army had both failed. He commented philosophically that they had each tried to ignore or to abolish one of the two fundamental bases of South Sulawesi society: social ranking and Islam. He now understood that both were too deeply entrenched to be overthrown by revolutionary means. Islamic practice could best be rectified through proper education and by providing a good example. As formal education had come to replace noble descent as the principal path to social status, the attack on feudal institutions had become less pressing.

By the 1980s, the royal ancestor cults had become almost exclusively the preserve of women of noble rank who had little modern schooling. Former militants such as Hakim were able to dismiss such practices as the product of backward superstition, *kepercayaan*, that would disappear as a better-educated generation came along, or as expressions of harmless local customs, *adat-istiadat*, and culture, *kebudayaan*. There are indications that they had made similar distinctions even at the height of the insurrection. Even when they could be identified as a source of violent resistance to the movement, the inhabitants of traditionalist villages such as Tana Towa, Amparita, and Cerekang remained relatively unmolested by Darul Islam militants. The surprising tolerance of the Islamic militants for local symbolic knowledge indicates that they always valued this sort of knowledge, insofar as it could be seen as complementary to cosmopolitan religious knowledge.

Somewhat paradoxically, former Darul Islam militants often appeared more tolerant of non-Islamic rituals, such as those in honor of the royal ancestor spirits, than they did of traditional Islamic rituals. Most remained firmly opposed to specific Islamic practices they regarded as *bid'a* or *shirk*. The most noteworthy among these was that they tried to prevent people from visiting the tombs of *shaikhs* and they would not enter a house in which the *Barasanji* was being recited. Since both practices are central to traditional weddings and funerals, this caused a certain amount of disruption every time a life-cycle ritual was performed. It was precisely because these practices claimed to be part of religion that modernists felt they had to draw a clear line between them and what they regarded as in keeping with fundamental Islamic teachings. Given their emphasis on the centrality of the *shariah* law to their definition of an Islamic society, they also found it difficult to accept the Dutch-inspired notion of a secular state.

These modernist attitudes present a marked contrast to the ones adopted by Panre Abeng and Haji Gama between 1890 and 1950. For them, the veneration of non-Islamic ancestors was seen as idolatrous, while the veneration of Islamic prophets and *shaikhs* was seen as the essence of piety. It was precisely because the ritual practices involved were so similar that Islamic traditionalists had to draw a clear line between the veneration of the royal ancestor, Karaeng Mamampang, and the veneration of the village *shaikh*, Bakka' Tera'. Given their acceptance of the separation of religion and politics, they saw no problem with enforcing their views of proper ritual behavior within the religious sphere while carrying out the policies of the colonial government in the political sphere.

Chapter 8

Official Islam and the Developmental State, 1965–2004

During the thirty-three years he was in power, President Suharto made use of all the symbolic models discussed in earlier chapters to legitimate his rule. This shows that the relevance of these models was not confined to South Sulawesi, but had an Indonesia-wide appeal. Suharto often boasted of his close ties to the traditional and charismatic authority of the central Javanese sultans of Yogyakarta, Surakarta, and Mangkunegara (chapter 2). In the official biography published by Roeder in 1969, Suharto's early life contains many parallels to the lives of Islamic heroes such as Shaikh Yusuf and Datu Museng who were also poor orphans that succeeded in marrying royal wives despite their obscure paternity (chapters 3 and 4). After his parents separated, Suharto's father placed him in the care of a second, a third, and a fourth foster mother. This part of the narrative recalls the childhood of the Prophet Muhammad as articulated in the *Barasanji* and of Andi' Patunru in the *Epic of the Three Boats* (chapter 5). In 1989 he made a well-publicized *hajj* to Mecca and returned with a new name, Muhammad, and an Islamic identity more in keeping with contemporary cosmopolitan norms (chapter 6). Roeder is careful to note that while he studied traditional Islamic mysticism with a village *kyai*, he also attended a modernist Muhammadiyah school for a while (chapter 7).

Despite these appeals to the charismatic authority of Islam, Suharto was at heart a bureaucrat who modeled himself on the Dutch colonial officers he had known in his youth, and especially of the governor generals who exercised supreme power from Batavia/Jakarta. In many ways, Suharto's New Order regime brought the policies of the late colonial state to completion. It encouraged the preservation of carefully sanitized versions of local traditions. It mandated the teaching of officially sanctioned versions of Islam in the public and private school systems. And it turned modern political activities like national elections into carefully orchestrated bureaucratic rituals. The main agencies of national development at the village level were the Village Cooperative Units (*Kooperasi Unit Desa*, KUD), which aimed at transforming the local economy; the family planning service (*Keluarga Berencana*, KB), which aimed at transforming the local kinship system; and the schools, which aimed at transforming local knowledge. Between 1965 and 1990, literacy jumped from 40 percent to 90 percent. Between 1970 and 2000, high school graduation rates went from 4 percent to 30 percent (Hefner 2000: 119–120). By 1988, every village had several kindergarten and elementary schools, every regency had several junior and senior high schools, and every province had a teachers' training college (IKIP). The Universitas Hasanuddin in Ujung Pandang served as the top of the educational system for the whole of eastern Indonesia. Scholastic achievement provided entry into the finely graded government bureaucracy.

The formation of an Indonesian national culture influenced the stance local actors in Ara adopted toward the symbolic and political models bequeathed them by their own history. Many of my closest informants occupied the lowest levels of the New Order bureaucracy, serving as schoolteachers or as civil servants in the Department of Education and Culture, *Departemen Pendidikan dan Kebudayaan* (*DepDikBud*). Education had become so essential to success that people in Ara often remarked that feudal privilege had finally been undermined by the fact that commoners were more motivated to achieve high academic rank than nobles were. In 1988, it sometimes seemed as if the titles that came from academic achievement (B.A., *doktorandus, doktor*) had more prestige than the titles that derived from noble birth (*andi', daeng, karaeng*). As members of the *pemerintahan*, government administration, civil servants saw themselves as agents of social transformation and viewed it as their duty to educate and uplift the *rakyat*, the great mass of unenlightened people. They were keen to converse with me about international politics, the latest scientific research, and the measures required to develop the national economy.

In the 1990s, leading Muslim intellectuals from both the Left and the Right began to articulate a popular and liberal critique of Suharto's corrupt regime, undermining his charismatic authority. The economic crisis of 1997 undermined his bureaucratic authority. The very success of his drive to modernize the country had undermined the appeal of his promotion of symbols of traditional authority. By the time his regime finally collapsed in 1998, Suharto had lost almost all of his legitimacy. At the beginning of the twenty-first century, political discourse in Indonesia was dominated by a vigorous debate over how best to reconcile Islamic law with electoral democracy, national unity, and provincial autonomy.

The Origins of Suharto and the New Order, 1921–1971

Several times the Commander [Suharto] himself went on secret missions to Dutch-held Jogjakarta. Dressed as a peasant, he took vegetables to the Sultan's kitchen. After a haircut, and a bath, not to mention a few drops of perfume, and properly dressed, Soeharto was led by the chief-cook Hendrobujono to His Highness Sri Sultan Hamengku Buwono IX. There were long conferences between the scion of a noble family and the son of a peasant—the two men united in their love for freedom. The sultan, despite Dutch pressure for cooperation, remained faithful to the Republic. Enjoying the respect of the masses and the leaders, Sri Sultan was the heart of the national resistance. Lieutenant Colonel Soeharto was his trusted field commander. (Roeder 1970: 124)

At the time he came to power in 1965, Suharto was almost unknown to the outside world. His authorized biography, *The Smiling General*, was first published in 1969 (Roeder 1970). In this text, Suharto claims that he was exposed in his youth to traditional Javanese notions of kingship, to charismatic Islamic notions of mystical knowledge, and to the bureaucratic forms of knowledge and power employed by the Dutch and Japanese colonial armies. He later drew on all these systems to bolster his political legitimacy.

Suharto was born in 1921 to a peasant family in Kemusu, a village near the old royal center of Yogyakarta in central Java. His father, Kartoredjo, was a somewhat mysterious figure. An article in a popular magazine published in 1974 suggested that Suharto might be the illegitimate son of Padmodipuro, an aristocratic descendent of Sultan Hamengkubuwono II. Other rumors even claimed that he was the illegitimate son of Sultan Hamengkubuwono VII. In 1974, Suharto called a press conference to angrily denounce all these rumors. "Suharto made an extended address to the hundred or so domestic and foreign journalists and senior officials in attendance . . . he presented to the press a bevy of aged relatives and acquaintances who could testify to the truth of what he said" (Elson 2001: 3; McDonald 1980: 9). Such protests did little to dampen the speculation, however.

Whatever the truth of his genetic paternity might be, it was Suharto's legal father, Kartoredjo, who oversaw his upbringing, albeit from a distance. Kartoredjo separated from Suharto's mother when he was two years old. After this break, Kartoredjo never cared for Suharto personally again, but saw to it that he was placed with a series of caretakers. He was first placed in the house of Kartoredjo's mother, who had been the midwife at Suharto's birth. Her niece, Amat Idris, served as his wet nurse. Suharto was four years old when his birth mother remarried. Kartoredjo then returned him to her care. At age seven, he started school in the village of Tiwir, four miles from his house.

At age nine, Kartoredjo removed Suharto from his birth mother for a second time and placed him in the care of his sister in Surakarta (Solo). In the same year, this sister and her husband, Prawirowiardjo, moved to the village of Wurjantoro near Wonogiri, where Suharto continued his elementary education. Prawirowiardjo was a typical lower-level *priyayi* functionary in the Dutch colonial government. Unlike many poor boys of his generation, Suharto thus had an early exposure to the modern disciplinary methods of the colonial bureaucracy and of the school system. In Wujantoro, Suharto also attended a Koranic school in the afternoon and joined the Hisbulwathan, an Islamic youth movement that exposed him to the teachings of Islamic modernism.

After Suharto completed elementary school, he went to live with one of Prawirowiardjo's sons in Selogiri in order to attend a middle school. When he was fifteen, he moved to the house of Hardjowijono, a friend of his father who lived in the city of Wonogiri. There he came under the influence of a *kyai*, Darjatmo, who instructed him in traditional Javanese-Islamic mysticism. Because he could not afford a school uniform, however, Suharto was forced to return to his mother's house in Kemusu before finishing middle school in Wonogiri. From Kemusu, he bicycled every day to a modernist Islamic middle school in Yogyakarta run by the Muhammadiyah, where he finished his studies at the age of eighteen.

After graduation, Suharto worked for a rural bank for a year before joining the Dutch colonial army (KNIL). In 1942 he joined the Japanese colonial police and then transferred to the Japanese-run "Self Defense Force" (PETA), receiving rigorous military training. After the Japanese were defeated in 1945 and Indonesian nationalists declared their independence from the Netherlands, Suharto joined the nationalist People's Security Corps. He was elected Deputy Commander of Battalion X, a unit stationed in Yogyakarta.

He achieved a feeling of stability and belonging in the ranks of these three forces. Their hierarchical organization and discipline were another source of inspiration in his later life.

In 1947, his foster mother, Mbah Prawirowiardjo, arranged his marriage to Siti Hartinah. She was the daughter of Raden Mas Ngabaei Sumoharjomo, a fifth-generation descendant of Mangkunegara II (r. 1796–1835). This placed her just within the boundaries of the Javanese nobility. Like Suharto's foster father, Sumoharjomo was a typical *priyayi* employee of the colonial state. Genealogists later determined that Suharto could trace his own descent back to Sultan Hamengkubuwana V of Yogyakarta (r. 1792–1828) through seven generations on his mother's side, and to Pakubuwana VII of Surakarta (r. 1830–1858) on his father's side.

On December 19, 1948, the Dutch attacked the headquarters of the provisional republican government in Yogyakarta and arrested President Sukarno, Vice President Hatta, and some ministers. Suharto retreated with his men into the surrounding villages, leaving his wife behind in the royal palace under the care of the sultan of Yogyakarta. This allowed him to cultivate his association with the royal court during the next few months. According to Roeder, Suharto and the sultan hatched a plan to retake Yogyakarta from the Dutch on March 1, 1949. Although they held the city for only a day, their success was an important sign of the continuing vitality of the resistance and strengthened Sukarno's hand in negotiations with the Dutch. For Suharto, the close association he cultivated with the sultan of Yogyakarta during these years helped to consolidate his connection to Java's royal traditions.

Following independence, Suharto was sent to South Sulawesi to restore order after a certain Captain Andi Aziz seized Makassar with the help of Indonesian troops who had belonged to the Dutch colonial army (KNIL). No sooner had Suharto talked them down than another rebellion led by the irregular "People's Army" of Arief Radhi had to be put down by force. Army headquarters now sent a Bugis commander, Kahar Muzakkar, back from Java to Sulawesi to take Suharto's place, thinking there would be less hostility to him than to a Javanese. Suharto advised against this measure, having observed Muzakkar in action against the Dutch in Java, but was overruled. When Muzakkar himself led an insurrection against the national government two years later, Suharto was proved right (see chapter 7). Suharto continued his slow rise through the ranks during the 1950s, and in 1963 he was put in charge of the Army Strategic Reserve Command (KOSTRAD) with the rank of Brigadier General. It was from this position that he put down the coup attempt of 1965 and began to consolidate state power in his own hands (Roeder 1970: 131–136).

By the beginning of the 1960s, the Communist Party of Indonesia (PKI) had acquired unprecedented influence over national politics through its alliance with President Sukarno. In 1963 the Party overplayed its hand by launching a land reform campaign that dispro-portionately threatened the interests of traditionalist Muslim clerics in Java, many of whom had accumulated large areas of land through their control of pious endowments (*waqaf*) and a general ethic of hard work and savings. The organization that represented these rural clerics, the Nahdlatul Ulama, cemented an alliance with elite factions in the military who opposed the PKI for their own reasons. Tensions within the villages were thus exacerbated by the appearance of vertical cleavages within the state itself. Competing factions within the state looked for and found allies in the larger society (Hefner 2000: 53–55).

On September 30, 1965, there was an attempted coup in Jakarta. Elements in the army associated with the PKI attempted to kidnap Generals Nasution, Yani, Parman, and four others. Nasution escaped; Yani and two other generals were killed while resisting; the other three were murdered while in custody. General Suharto took command of the loyal military units in the capital, and by October 2 the coup was essentially over. Suharto quickly marginalized President Sukarno, who was compromised in the eyes of many senior military officers by his close ties to some of the coup leaders and to the PKI. Suharto portrayed the coup attempt as a Communist plot. Conservative Islamic groups that had watched the growth in power and membership of the PKI with mounting fear launched a wave of massacres that were especially horrific in Java and Bali. As many as 500,000 people associated with leftist political parties are believed to have died in late 1965 and early 1966. In South Sulawesi, some killings occurred in the agrarian heartland of Bone, but on a much smaller scale that in Java.

Having marginalized Sukarno, Suharto declared a New Order in which the conflicts between the PKI and the militant Islamic parties would be replaced by an orderly process of national development. The goal of the New Order was to develop Indonesia's economy and society in a rational, bureaucratic manner that suppressed all religious and political conflict. *Bapak* Suharto, Father Suharto, played a central symbolic role in New Order ideology, along with his wife, who became known as *Ibu* Tien, Mother Tien. State ideology continued to be based on the *Pancasila*, the Five Principles enunciated by Sukarno on June 1, 1945. These were belief in a "singular" God, nationalism, humanitarianism, social justice, and democracy. At the time they were formulated, Sukarno was attempting to finesse the disagreements among Islamic, secular, and socialist tendencies in the nationalist movement. The studied ambiguity of the five principles continued to provide a useful umbrella under which the central government could foster the development of a national culture (Hefner 2000: 41).

Although the army had used the Islamic Right to neutralize the Left in the massacres of 1965–1966, senior officers such as Suharto continued to view political Islam with suspicion. They had spent much of the 1950s fighting the Darul Islam movement in Sumatra and Sulawesi and continued to doubt their loyalty to the nation's center. Suharto saw to it that Islamic organizations were depoliticized. But in a concession to religious feeling that was to have political consequences later on, Suharto made religious education compulsory from elementary school through university. Between 1967 and 1971, the Department of Religion was placed under the control of traditionalist Javanese Muslims belonging to the Nahdlatul Ulama. The staff of the department grew by 60 percent during this period (Hefner 2000: 119–120).

National Elections as Traditional Rituals, 1971–1987

In the early years of his regime, Suharto often claimed a measure of traditional authority by alluding to the fact that he could claim remote genealogical ties to the two most prestigious royal courts of central Java, Yogyakarta and Surakarta. Although he was born to a humble family, after acquiring religious knowledge and military success he was able to marry a noble woman who traced her descent from the royal house of Mangkunegoro. During the war of national liberation, Suharto had gained the trust of the most prestigious hereditary ruler in all of Indonesia at the time, Sultan Hamengkubuwono IX of Yogyakarta. As his hold on power stabilized, Suharto and his wife drew on their ties to the courts of central Java to represent themselves as the metaphorical *Bapak* (Father) and *Ibu* (Mother) of the nation. Their marriage could be portrayed as a symbolic reunification of Yogyakarta, Surakarta, and Mangkunegoro, royal houses that had been created by VOC meddling during the eighteenth century.

In 1971, Suharto attempted to place the legitimacy of his rule on a sounder footing by organizing national elections. In the run-up to these elections, he transformed the army's "functional groupings" *golongan karya*, into a political party, Golkar, and pressured all government employees into joining it. Golkar won about 62 percent of the vote. In 1973, Suharto implemented a "simplification" of the party system. All the non-Muslim parties were brought together as the *Partai Demokrat Indonesia* (PDI). The Muslim parties were forced together by Suharto into the *Partai Persatuan dan Pembangunan*, the Unity and Development Party (PPP). When national elections were again held in May, 1977, Golkar again won about 62 percent of the vote.

In a paper published in 1980, Schulte Nordholt argued that the Suharto regime had enhanced its appeal to the Javanese masses by reinterpreting Sukarno's *Pancasila* ideology in terms of the ancient Javanese *moncopat* symbolic system. According to this system, the human body, local villages, and royal polities are all thought of as being composed of four outlying parts oriented to the cardinal directions, with a fifth unit in the center to unify them. While inspired by the structuralism of the Leiden school, Schulte Nordholt avoided the circular reasoning found in many of their analyses. He was able to show that key political actors had manipulated the Pancasila quite consciously. For example, Professor Notonagoro of Gadjah Mada University delivered a lecture in 1967, in which he clearly stated that the five principles were a legacy of the *moncopat* system. This interpretation formed the basis of one of Suharto's first decrees, a presidential instruction on the *Pancasila* issued in 1968, and of a little black book that all military personnel had to carry in their pockets.

Another example was provided by a lecture delivered in October 1977 to a conference of Christian students by General Widodo. The general described the *Pancasila* as the ancestral element that binds together the diverse religions of Indonesia (Islam, Catholicism, Protestantism, Hinduism, and Buddhism), much as the child unites the "four siblings" that accompany it at birth (the amniotic fluid, the womb, and the milk from each breast). General Widodo saw an early version of Schulte Nordholt's article and approved of his general interpretation (Schulte Nordholt 1980).

In the elections of 1971 and 1977, the Islamic PPP was made to represent the first principle, belief in God, and the nationalist PDI the fifth principle, social justice. Golkar then represented the three unifying principles at the center: nationalism, humanitarianism, and democracy. This enabled Suharto to both recognize and marginalize dissenting political opinions by symbolically locating them at the margins of the government party.

John Pemberton attacked Schulte Nordholt's appeal to "traditional" Javanese culture, claiming that the very idea of "Java" had only emerged during the colonial era when all things Javanese had been reified as a monolithic "culture" in contrast to all things Dutch. In its search for a Javanese cultural essence, subsequent anthropological analysis had only contributed to this reification. Pemberton argued that Suharto's attempt to treat national elections as "traditional" rituals was intended to depoliticize Indonesian society in much the same way that the Dutch had intended to depoliticize it during the 1930s. In both cases, a sanitized version of local *adat*, custom, and *kebudayaan*, "culture" was promoted across Indonesia as a buttress against the revolutionary forces of socialism and of Islamic modernism (Pemberton 1994: for the application of this policy in South Sulawesi see Gibson 2005: 181–186).

In the national elections of 1982, Golkar won 64 percent of the vote nationally. Pemberton observed these elections in the central Javanese city of Solo and despaired of finding any real resistance to Suharto in the formal political sphere. Pemberton claimed that the last remnants of resistance to Suharto's reified view culture could be found in the periods of disorder, *rebut*, that occurred in authentic village rituals. He argued that ritual disorder was the only way left to peasants during the New Order to dissent from the seamless hegemony of colonial Javanese culture. This form of dissent was a completely inarticulate, marginal, and chaotic form of behavior. The regime recognized the subversive potential of these periods of disorder and attempted to ban them.

One of the problems with Pemberton's analysis was that it neglected the long line of writers who have noted that periods of disorder in the middle, liminal phase of a ritual are extremely common and that they typically serve to reinforce traditional authority, not to subvert it (van Gennep 1909). Gluckman and Turner argued that such disorder allowed the periodic release of social tensions (Gluckman 1963; Turner 1969). Bloch has argued that disorder provides such a horrifying image of what would happen if age, rank, and gender roles broke down altogether that participants are happy to see the traditional moral order restored at the end (Bloch 1986).

Another problem is that Pemberton left no real room for meaningful political resistance, much less for the true sources of revolutionary political change. In his single-minded critique of the concepts of culture and tradition, Pemberton entirely neglected the existence of Islamic and nationalist models of the state. It is almost as if Pemberton actually shared the view of the Leiden anthropologists he spends so much time criticizing in that both view "cultures" as forming unified, coherent wholes. It is only if cultures are coherent in this way that people would be unable to put their criticisms of the existing order into words. As we have seen, however, nothing could be further from the truth in Indonesia, where centuries of writing and of exposure to global currents of opinion from all over the Islamic world have provided local actors with any number of competing models of the ideal political and religious order.

It is only if one assumes that people have access to only one ideal order that one will have to search for mute, inarticulate forms of resistance. This view of resistance recalls that of James Scott, who saw the essence of political dissent in the inarticulate foot-dragging of Malay peasants rather than in the Islamic prophecies that pervaded village life (Scott 1985). What both authors ignore is that Malays and Indonesians have been carrying on highly differentiated and articulate struggles against European domination for centuries in the name of the charismatic authority of Islam. Hefner notes that by the time Pemberton published his analysis of the 1982 elections in 1994, it was already clear that Suharto's support of "Javanese tradition" had been only a temporary expedient, quickly superseded by a turn toward Islam (Hefner 2000: 248 n.66). Political scientists had known since the middle of the 1980s that the major players in Indonesian politics were going to be the leaders of large Islamic organizations such as the Muhammadiyah and the Nahdlatul Ulama. It was only anthropologists, with their excessive attention to local forms of knowledge, who missed this (Gibson 2000).

In 1983, Suharto persuaded Abdurrahman Wahid to withdraw the Nahdlatul Ulama from party politics. In the national elections of 1987, Golkar's share of the national vote rose to 73 percent. Support for the Islamic PPP fell to just 16 percent since it now represented only modernist Muslims. Wahid was rewarded by Suharto with an appointment to the People's Consultative Assembly as a Golkar delegate. The nationalist PDI share rose to 11 percent. Since the military appointed 100 out of 500 delegates and Golkar controlled another 299 seats, just 101 seats were left for the opposition (Hefner 2000: 167–168).

The Development State and the *Hajj* in Ara during the New Order, 1965–1989

In chapter 6 we saw that the suppression of piracy and the introduction of steam technology led to a wave of economic "globalization" as the export of bulk produce such as copra from the tropics to Europe became profitable. This led to a wave of religious globalization as the coconut growers of Selayar used their profits to perform the *hajj*. In the 1970s and 1980s, the male boat builders and female cloth merchants of Ara experienced a similar wave of economic prosperity as they profited from Indonesia's general economic development. They too used their profits to finance the performance of the *hajj* on an unprecedented scale.

Throughout the nineteenth century, Lemo Lemo and Bira were far more prosperous than Tanaberu and Ara. Tanaberu lay on an exposed beach and was periodically devastated during times of war. It was last burnt to the ground in 1954 during the Darul Islam period. Historically, the men of Bira specialized in navigation and sailing. The most successful became merchants who financed the building of boats. For generations, the merchants of Bira had a monopoly on local capital and the boat builders of Ara were wholly dependent on them for orders. Collins gave a vivid account of the tense relations between the rich merchants of Bira and the poor artisans of Ara in the 1930s (Collins 1936, 1937).

By the 1960s, however, Lemo Lemo and Bira had fallen on hard times. Most of the local hardwood forests had been cut down and the boat builders of Lemo Lemo had to move to Tanaberu where timber could be brought in by truck to the beach. The original settlement of Lemo Lemo was almost entirely abandoned. The merchants of Bira were increasingly unable to finance the building of large cargo boats because they could not afford to install the engines that were needed to supplement the sails.

Between 1953 and 1965, many of the inhabitants of Ara fled to Flores, Irian Jaya, and Sumatra to escape the food shortages and violence caused by the Darul Islam rebellion. There they discovered new sources of wood and new markets for their boats. As Suharto's regime restored order throughout Indonesia in the late 1960s, the villagers of Ara used these resources to become more prosperous than the villagers of Lemo Lemo and Bira. By 1988, the men of Ara had acquired a national reputation for their expertise in building the largest wooden sailing boats and could demand a premium wage for their labor. Only ethnic Chinese entrepreneurs could afford to install the engines on the largest boats and they were responsible for most of the new commissions.

Women also contributed to Ara's new prosperity. Migrant boat builders spent up to nine months a year away from home, and brought back most of their wages in a lump sum. Their wives used their wages to buy sewing machines with which they embroidered cushion covers and tablecloths. They found that the sale of these articles earned them enough to feed their children while their husbands were away. In subsequent years, many men were then able to invest their wages in small vans and trucks that plied a daily route between the city of Ujung Pandang and the villages of Bulukumba.

Many older women became peripatetic cloth merchants after the deaths of their husbands. Successful business men and women redistributed a significant proportion of their wealth by sponsoring elaborate Islamic rituals held on the occasion of the first trimming of a child's hair, circumcision, marriage, and death. During these rituals, traditionalist *imams* were hired to recite the *Barasanji* and *dhikr* over elaborate offerings of food and drink. They would then lead the participants to the tombs of Datu Tiro, Bakka' Tera, and Pua' Janggo' to make vows, *nazar*.

The most successful merchants often capped their careers by performing the *hajj*. Participation in the *hajj* was limited under Sukarno to those who won a government lottery. Only two couples from Ara were able to make the pilgrimage between 1950 and 1970. In the five-year plan that began in April, 1969 Suharto abolished the lottery and allowed everyone who could afford it to go on the *hajj*. After the rules changed, Ara averaged eight *hajjis* each year. The *kepala desa* in 1988, Daeng Pasau, said that according to his calculations over 170 people from Ara had made the pilgrimage over the previous 20 years. This is an extraordinarily high figure by any standards, but particularly remarkable in comparison to neighboring villages. Ara set a record in 1988 when it sent twenty people to Mecca. By contrast, Bira sent one in 1988 and none in 1989. Tiro sent none in 1989, and Tana Lemo (the *kelurahan* formed by the fusion of Tanaberu and Lemo Lemo) sent only two. Each of these three villages had a much greater population than Ara.

In 1989, the *hajjis* from Ara included three married couples and three widows, each of whom had accumulated the necessary $3,000 through their own business acumen. The pilgrims were organized into *ragu* of ten people plus a leader, *ketua*, who received Rp 60,000 in compensation. Five *ragu* were grouped into a *kelompok* of fifty-five individuals. Ara thus had almost enough *hajjis* in 1989 to form its own *ragu*, and only had to include two more people from Tanette in northern Bulukumba to make up a full set.

The main worry of pilgrims seemed to be about getting separated from their group and left behind in a land where no one spoke Indonesian. During the 1989 trip, the *kelompok* leader himself got lost for twenty-four hours. Villagers also told a cautionary story about one of the pilgrims from Tana Lemo. He returned home several weeks early because he could not stand the heat in Saudi Arabia. His whole body broke out in a rash and he felt his skin was on fire. Since no one else in his group was affected, it was rumored that his affliction was due to divine anger at his habit of shortchanging customers at his filling station.

But no one denied it was hot in the Holy Land: they said that 144 Indonesians had died that year from the heat.

In 1989, I attended the ceremony to welcome the *hajjis* back, which is held in front of the *bupati's* office in Bulukumba every year. The relatives of Ara's nine *hajjis* chartered four minibuses to take them to Bulukumba. At 5 p.m. on August 1, ten large buses roared in from Ujung Pandang and parked in the middle of a large field that had been surrounded by barbed wire and police to keep the crowd back. Around 250 pilgrims descended to sit in rows of chairs and listen to a speech by the *bupati*. When the short ceremony was over and it was announced that the *hajjis* could begin collecting their bags, there was no holding the crowd who poured onto the field to greet their returning relatives with great emotion. By 7 p.m. everyone was back in the buses and by 9 p.m. we were all back in Ara. There the whole village had turned out to greet them, in the state of high excitement generated only by large crowds. The pilgrims went first to the mosque to say a prayer of thanks for their safe return, and then home to rest.

In the 1980s most of Ara's *hajjis* were men and women with little formal education that had spent most of their lives life as carpenters and seamstresses. They used their savings to become merchants and their profits to perform the *hajj*. For such people, Islam continued to form an alternative source of prestige to the one provided by the traditional system of hereditary ranks and to the modern system provided by formal schooling.

The Local Elections of 1989 as a Bureaucratic Ritual

Suharto occasionally appealed to ancient Indonesian traditions of kingship, but he relied much more systematically on the disciplinary practices of the modern civil and military bureaucracies. He grew up in the household of a functionary in the Dutch civil service, studied at Dutch colonial schools, and was trained by the Dutch colonial army. He received further police and military training from the Japanese, who stressed implicit obedience to superiors even more strongly than did the Dutch. Suharto organized the army, the police, and the civil administration as three parallel bureaucracies each of which maintained a separate system of surveillance over every level of society. He incorporated all local civil servants into the state party, Golkar, and made them instruments of his bureaucratic vision of society.

As Pemberton noted in Java in 1982, state-sponsored ritual took the place of authentic political debate in South Sulawesi in 1988. I was far less impressed either by the Javanese traditionalism of the regime or by its promotion of Islamic piety than I was by its devotion to a cult of militaristic development. Schoolteachers and civil servants had to cultivate endless patience and humility in the face of their bureaucratic superiors. Much of a civil servant's life was taken up attending an endless sequence of meetings and ceremonies. The nightly domestic news on television was mostly taken up by pictures of civil servants attending meetings and ceremonies and struggling valiantly to stay awake. While very little seemed to get done on these occasions, something very significant was being communicated both to those who participated and to those who did not.

In the late 1980s, the entire population was periodically drawn into state ceremonies, many of which were organized around the school system. Students all across Indonesia had to don their uniforms and perform martial exercises on their village fields on the seventeenth day of every month. Especially elaborate celebrations were performed annually on August 17. These were to commemorate both the declaration of independence on August 17, 1945 and the salvation of the nation from the communist menace on September 30, 1965. In the ideology of Suharto's New Order, both events were of equal importance in the history of the nation.

An image of state power was projected during such ceremonies, which emphasized a hierarchy of political ranks that was uniform throughout the nation, from Irian in the east to Sumatra in the west, from Java in the south to Sabah in the north. Every official belonged to the same series of political units, from the *dusun, desa, kecamatan, kabupaten,* and *propinsi* to the nation as a whole. Every official put on the same uniform and performed the same drill at the same time throughout the country, commemorated the same national heroes, and celebrated the defeat of those who questioned the state ideology, particularly the atheists.

Since anthropologists were prohibited from conducting fieldwork for the six months preceding and the six months following national elections, I was extremely fortunate to be in the field in 1989 when the regent of Bulukumba began a major reorganization of local government. Twenty-seven new *desa*, villages, or *kelurahan*, urban wards, were created. Bira was divided into two *desa*. The regent decided it was time to hold an election to replace the *kepala desa* of Ara, Daeng Pasau.

The election was hotly contested by four candidates and long-suppressed factional rivalries were allowed an outlet for the first time since 1967. The front-runner was Daeng Pasau's brother, Haji Mustari.

Haji Mustari had won the last real election in 1967, which had been the first local elections held since the beginning of the Darul Islam rebellion in 1952. When Suharto pressured all civil servants to join Golkar in 1971, Haji Mustari resigned from office and moved to Bulukumba where he opened a construction business. He was a former Darul Islam schoolteacher and a firm believer in the modernist doctrines of the Muhammadiyah. He joined the Islamic PPP when it was organized in 1973.

While working as a building contractor in Bulukumba City during the 1980s, Haji Mustari discovered that he could not survive even in private business without connections to the government party and so eventually joined Golkar. He then obtained a position in the government-run development agency (KUD) that extended credit, mostly to farmers. Controlling this agency at the village level gave him the ability to distribute a significant amount patronage. As there was very little farming in Ara, Haji Mustari worked to get loans for women to purchase materials for embroidery and for retired boat builders to make models for tourists. These artisans formed one wing of his support. Another wing of his support consisted of former Islamic militants such as my host, Abdul Hakim, and his sister's husband, Muhammad Idris.

Opposition to Haji Mustari centered on the private merchants of the bazaar. The government loans controlled by Haji Mustari charged an interest rate of only 1 percent per month, while the private merchants were charging up to 30 percent a month. They thus viewed the KUD as a threat to their financial interests. They began quietly to organize their economically dependent clients to vote for Pahatte, a well-educated young man without strong views on religion.

Another group of opponents centered on the descendants of colonial officials of the interwar period. This group had managed to retain many salaried positions in the local government throughout the twentieth century, having been as ready to work with the Dutch and Japanese as with the Republican governments. They backed Haji Arifin, whose commoner father, Pantang Daeng Malaja, had been one of Gama's right-hand men in the 1920s and 1930s. Members of this group also saw themselves as fervent Muslims, but of a more traditional style than the modernist schoolteachers. They tended to support traditional Islamic rituals such as the recitation of the *Barasanji*.

One of Haji Gama's grandsons, Andi' Azikin, came forward as a fourth candidate. He seemed to have no independent power base, but regarded himself as a plausible candidate simply because of his descent from Haji Gama.

Realizing they had no chance of defeating Haji Mustari in a field of four, the bazaar merchants tried to persuade Haji Arifin to withdraw in favor of his nephew, Pahatte. When he refused, the merchants threw their bloc of votes to Andi' Azikin in protest. Haji Mustari won with 582 votes (35 percent); Haji Arifin came in second with 467 votes (28 percent); Andi Azikin came in third with 441 votes (27 percent); Pahatte came in fourth with 94 votes (6 percent); and 67 votes were spoiled (4 percent). Had Haji Arifin withdrawn, then all of his votes and half of Andi' Azikin's votes might have gone to Pahatte, who would then have defeated Haji Mustari by 200 votes.

Haji Mustari was a well-educated modernist Muslim whose family had been prominent in village politics for over sixty years and who had access to state patronage. It was the combination of these characteristics that allowed him to forge a winning coalition. The outcome of this election can only be understood by reference to the short-term strategies employed by competing economic interest groups, the medium-term strategies employed by individuals and families to climb the social ladder, and the long-term strategies embedded in religious models. Attempting to explain the event in terms of any one of these levels would impoverish our understanding of the whole.

Religious Education and Suharto's "Islamic Turn," 1988–1999

While the national political system appeared completely frozen in 1988, strong political and religious currents continued to run just beneath the surface. Indonesia's economic development during the 1970s and 1980s enabled villagers everywhere to finance a more ostentatiously pious way of life. Elaborate mosques equipped with loudspeakers to announce the call to prayer were constructed; radios and televisions, which carried religious broadcasts influenced by Egyptian and Saudi Arabian definitions of Islamic orthodoxy, were purchased; and the *hajj* was performed by ever-growing numbers. Suharto increasingly found himself on the defensive in relation to these trends toward greater piety as it was defined in the Middle East.

While the older generation used its newfound wealth to finance an increase in traditional Islamic piety, the younger generation was learning a new form of official Islamic piety in the state and private school systems. By the 1990s, these schools had produced a whole generation of self-conscious Muslims without any formal ties to titled clerics, Islamic parties or Koranic schools. Attendance at state schools for higher Islamic education (IAIN) quadrupled to 100,000 between

1979 and 1991, or 18 percent of all students in higher education (Hefner 2000: 119–120). During the tenure of Munawir Sjadzali as minister of religion from 1983 to 1993, many postgraduate students were sent to the United States and Western Europe to study Islam instead of to the Middle East (Hefner 2000: 110).

A new generation of Islamic activists emerged during the 1990s. In the face of Suharto's repression, they abandoned the earlier modernist goal of Islamizing the state and turned toward the use of predication as a means of Islamizing Indonesian society. During the period of late colonial repression from 1927 to 1941, the Muhammadiyah had also withdrawn from open nationalist agitation and turned toward the reform of social and religious life. The younger modernists found their base in the new middle class of state-educated Muslims who found employment in the government and government-backed businesses. They argued that good Muslims had to avoid the "low politics" of political power and patronage, and had to develop a "high politics" emphasizing social justice and democracy.

Despite his ties to the courts of central Java and to the colonial bureaucracy, Suharto also stressed his early familiarity with traditional Javanese mysticism. As his hold on the army began to slip in later years, he turned to Wahhabi-influenced Islamic modernism as a new source of religious legitimacy.

> [The] president's range of spiritual interests underwent an important change in the early 1980s. While continuing his mystical exercises, the president hired a Muslim preacher previously active in the armed forces to serve as his personal instructor in Islamic devotion. . . . As news of the president's piety spread in the early 1990s, many pro-government Muslims pointed out that in his youth, Soeharto had briefly attended Muhammadiyah schools. . . . (Hefner 2000: 83)

Suharto's self-transformation was complete when he and his wife performed the *hajj* in 1991. At the end of this trip, they accepted the new names of Muhammad and Fatimah from King Fahd of Saudi Arabia, much as the sultans of Banten had sought recognition from the *sharif* of Mecca in the seventeenth century.

Suharto recognized the potential importance of the younger generation of Muslim intellectuals. Beginning in 1990 he tried to co-opt it for his regime by agreeing to sponsor the Association of Indonesian Muslim Intellectuals, ICMI. He hoped to use this organization against those in the army who were becoming increasingly critical of the growing power of his cronies in the state sector of the economy. ICMI brought together a diverse array of individuals, whom Hefner groups together as government bureaucrats, political activists, and independent intellectuals. Official positions in the organization were dominated by Western-educated government bureaucrats such as the Minister of the Environment Salim, a Berkeley-educated economist, and the German-educated engineer Bacharuddin Yusuf Habibie.

Habibie was born in the South Sulawesi port of Pare Pare in 1936, the son of a Bugis father from Gorontalo in Central Sulawesi, and of a Javanese mother from Yogyakarta. Suharto first met Habibie in 1950 during the time he was posted to South Sulawesi and the two formed a lifelong friendship. When Habibie's father died in 1950, his family moved to Bandung in west Java, where he later attended the Bandung Institute of Technology, before going off to Germany to study aeronautical engineering. Habibie returned to Indonesia in 1974 when Suharto asked him to lead Indonesia's State Aircraft Industry. In 1978 he was made minister of technology.

These government elites saw ICMI as a way of increasing the presence of Muslims in the state and the private sector in opposition to the disproportionate political and economic power of Christians and ethnic Chinese. At first it was assumed that the organization was only a scheme to shore up the Islamic vote for the government party in the 1992 elections, but when 300 ICMI members were appointed to the 1,000-member assembly in 1993, middle-level bureaucrats rushed to join. Pro- and anti-ICMI factions developed throughout the state and university systems, announcing the first open splits within the ruling elite since the 1960s.

An assortment of political activists also became involved in the organization. One group hoped to use Suharto to reduce military influence over the state and the economy. A second group was more conciliatory toward the military and advocated a more gradual reform of the New Order. A third group hoped to Islamize the government party, Golkar, and had no problem allying itself with the "Green" (Islamic) faction in the army to do so. Suharto sided with the reformist faction from 1990 to 1994, then turned to the ultra-conservative faction from 1994 to 1998. Another group active in ICMI consisted of independent intellectuals. While they seem to have played an important role in the first three years of the organization, independent scholars such as Nurcholish Madjid seem to have lost all influence by 1993.

Internally factionalized itself, ICMI also faced opposition from old school anti-Muslim generals in the army and from Muslim leaders who feared a return to the low politics of state patronage. An uneasy alliance developed between these generals and Abdurrahman Wahid, the leader of the NU. Wahid had withdrawn from the state-mandated

Muslim party in 1984 and supported the government party in the 1987 elections. But when Wahid refused to support ICMI in 1990, Suharto abruptly turned against him and launched a covert campaign to remove him from the leadership of the NU.

Fearing an alliance between the secular nationalist "reds" of the PDI under Megawati Sukarnoputri and the "greens" of the NU in 1994, Suharto worked to split both organizations by creating pseudo-factions within them. This campaign of dirty tricks had inconclusive results. In 1996 the regime turned to more extreme measures, mobilizing paramilitaries trained by Suharto's son-in-law General Prabowo. They fomented anti-Christian and anti-Chinese riots in late 1996 in an effort to distract attention from the growing calls for reform.

Meanwhile, the influence of Suharto's daughter, Tutut, was growing. She saw Habibie as her main rival to succeed her father. Tutut persuaded her father to distance himself from Habibie and ICMI. The green generals, including Suharto's son-in-law, Prabowo, remained loyal to Habibie. Tutut then made overtures to Wahid and in 1996 Wahid suddenly reconciled with Suharto again. Wahid openly blamed the violence on a faction in ICMI, not daring to name the ultraconservative generals who were really behind it.

In August, 1997, a major economic crisis hit Southeast Asia. Suharto's increasingly desperate attempts to stay in power came to an ignominious end on May 5, 1998 when he was forced to resign. His vice president, Habibie, took over as interim president. Habibie agreed to hold national elections in a year's time. In these elections, the secular nationalist *Partai Demokrasi Indonesia–Perjuanangan* (PDI-P), led by Sukarno's daughter Megawati, received 35.7 million votes and 153 seats in the people's assembly. The three largest Islamic parties won 32.1 million votes and 143 seats. Habibie's state party, Golkar, received only 23.7 million votes and 120 seats. In the negotiations that followed, a Muslim traditionalist, Abdurrahman Wahid, ended up as president; Sukarno's daughter, Megawati, as vice president; and a Muslim modernist, Amien Rais, as head of the People's Assembly. The triumph of liberal Islamic leaders was not an outcome many anthropologists would have predicted, focused as they tended to be on the most localized forms of symbolic knowledge in Indonesia.

In June, 2000, I visited Ara in the immediate aftermath of the national elections of 1999. I found that people were disappointed with their outcome. Abdurrahman Wahid was viewed as ill educated, irresponsible, and too physically handicapped by his near-blindness for the job of president. Habibie was regarded as a favorite son, having grown up in Pare Pare; as intelligent and well educated, having studied engineering in Germany; and as a devout Islamic modernist, having led ICMI. Habibie won about 60 percent of the vote in South Sulawesi as a whole, making it one of the few provinces in Indonesia that remained loyal to the party founded by Suharto.

The Makassar I spoke with were particularly critical of Wahid's policies on local autonomy. On the one hand, they saw his toleration of separatists in East Timor, Papua, and Aceh as threatening national unity and their own economic interests. Bugis and Makassar have been among the most mobile and enterprising ethnicities in Indonesia since the VOC wars of the seventeenth century. Their superior knowledge of market conditions and long-distance trading ties often allowed them to take advantage of subsistence-oriented peasants in places such as Papua, Kalimantan, and central Sulawesi. They perceived a strong national state and uniform code of law, preferably based on Islamic norms, as in their economic interest. Subsistence-based ethnic groups, on the other hand, enthusiastically embraced local autonomy at the level of the *kabupaten* level, and a revival of customary *adat* law, as the only way to safeguard their rights to their ancestral land from outsiders such as the Bugis and Makassar. The outbreak of violence between local peoples and immigrant groups from other parts of Indonesia threatened the livelihoods of many Bugis and Makassar, who began streaming back to South Sulawesi from Ambon and Central Sulawesi.

On the other hand, many Makassar perceived Wahid's cabinet as biased toward the Javanese center, following a pattern that went back to the late colonial period. In April, 2000, the one cabinet member from South Sulawesi, Jusuf Muhammad Kalla, was forced to resign over allegations of Corruption, Collusion, and Nepotism (KKN, one of the buzz acronyms of the day). He had been appointed minister of industry and trade in Wahid's first Cabinet of National Union (October 1999–August 2000). Abdul Hakim and others were of the opinion that Jusuf Kalla was innocent of wrongdoing and that he had been targeted for removal so that he could be replaced by another Javanese member of Wahid's inner circle, Luhut Pandjaitan. Thus they viewed Wahid and his cabinet as weak on national unity, Java-centric, backward on religious questions, and ignorant of modern science and technology.

In Ara, Abdul Hakim worked as one of Golkar's main organizers in Ara during the 1999 elections. One of his sons and one of his daughters supervised two of the four polling stations during the election. Muhammad Nasir, the reciter of the *Sinrili' Datu Museng*, acted on behalf of the united Islamic party that Suharto had created in 1973, the PPP. There were 1,013 ballots in Ara, of which 782 recorded a vote in the national election. Of these, 75.3 percent went to Golkar under the leadership of Habibie; 9.8 percent went to the PPP; only 4.6 percent went to the PDI-P under the leadership of Megawati; 3.6 percent went to the PAN under the leadership of Amien Rais; and 2.7 percent went to the Partai Nasional Indonesia-Front Marhaenis. The remaining thirty-one votes were distributed among ten other parties. Almost 90 percent of the votes were cast for Islamic parties if one counts Golkar under Habibie among them.

In 2000, my reading was that when Suharto's South Sulawesi protégé, Habibie, joined forces with the head of the Muhammadiyah to found ICMI in 1990; any lingering suspicion of Suharto's commitment to an Islamic social order was dispelled. Former Darul Islam guerillas such as Abdul Hakim who had resisted pressure to join Golkar in the 1970s became enthusiastic supporters in the 1990s. The people of Ara also remained supportive of Golkar because they had survived the economic crisis of the previous three years better than most Indonesians had. Their occupations remained relatively independent of the global economy of the 1990s and so were not directly impacted by the collapse of the rupiah in 1998. Most men continued to build boats using local materials for a national market and most women continued to produce embroidery for a local market.

Ara in 2000

After years as a village schoolteacher from 1960 to 1990, my host Abdul Hakim had finally done quite well in the 1990s. When Haji Mustari became *kepala desa* again in 1989, he handed over his control of the KUD to Abdul Hakim, who spent the next few years visiting the little colonies of Aran boat builders scattered across Kalimantan, Maluku, Irian Jaya, and Java. He prospered as never before. In 1999, Hakim was able to sponsor an elaborate wedding ceremony for two of his younger daughters, Subindi and Akira. He drew on the research we had conducted together in 1988 and 1989 to perform the most "complete," *lengkap,* set of wedding rituals the village had seen in a generation. They were recorded for posterity on a visual compact disc (VCD), a technology that had become widely available in the intervening decade.

Hakim increased his standing outside the village by volunteering to serve as an expert on the local customs of the *kebupaten* of Bulukumba. In the 1990s, the governor of the South Sulawesi created a *Taman Mini* or "Miniature Garden" outside Makassar to showcase the diverse cultures of the province. Each *kebupaten* was invited to build a traditional noble house in the garden. Hakim drew on research we had conducted together in the 1980s on the house built by *Gallarrang* Baso Sikiri of Ara in the 1880s, which was still standing in the 1980s. In 2004, Hakim was finally able to perform the *hajj*.

Hakim's close friend and fellow-modernist, Alimuddin, took over as village *imam* in the mid-1990s. He and Hakim disapproved of many traditional rituals that had once been suppressed but were being practiced openly again in 2000. Offerings were again being placed on the tomb of Bakka' Tera', the village Saint, every Thursday night. These rituals were organized by a great niece of *Gallarrang* Daeng Makkilo, the noble ruler of Ara who died in 1913. Dessibaji continued to recite the *Sinrili' Datu Museng*. The royal spirit medium, Haji Titi, continued to invoke Karaeng Mamampang in séances.

Hakim and Alimuddin disapproved of these practices, but they did not interfere with them. They said of this and many other "local customs" that these were the affairs of "the people," *rakyat*, and that it was not the business of members of "the government," *pemerintahan*, to interfere with them. The older generation of Islamic modernists had thus grown more and more tolerant of ritual pluralism as they aged. They saw it as their responsibility to keep the peace by ensuring that traditional rituals were performed discreetly so as not to antagonize well-educated members of the younger generation who were not so tolerant.

For some, like Abdul Hakim, the invocation of ancestor and nature spirits definitely remained unacceptable as contrary to religion. Elaborate life-cycle rituals, however, were acceptable as expressions of local culture, even though they too had been repressed in Darul Islam times as perpetuating feudal social relations. In particular, elaborate wedding rituals and huge marriage payments were seen quite rightly by the provisional Darul Islam government as the foundation of the hierarchical social relations that they were determined to replace by an Islamic egalitarianism. As we saw earlier, when Hakim got married in 1958 at the height of the Darul Islam insurrection, the ceremony was reduced to the Koranic minimum. The marriage payment was fixed at the lowest possible level suggested by the scriptures (Rp.125, equivalent to a few dollars at the time). The ceremony itself was limited to the signing of the *nikah* contract.

All this had changed by the 1980s. As Hakim put it, all sorts of old practices had "sprouted up" again. Hakim's eldest child, Nurhadi, married Mustari in the early 1980s. Mustari was a particularly devout modernist Muslim, and prayed in the mosque several times a day. But he was also from a very high-ranking family in the neighboring village of Caramming, and he had a *doctorandus* degree from the teacher's training college, IKIP. Hakim's oldest son, Ahmad Mahliadi, married the daughter of a local farmer in 1988 just before I arrived. Hakim said he spent 3 million rupiah (US $1,700 in 1989) on this wedding, more than twice the amount a high school teacher with a B.A. degree earned in a year. His house was still covered with the bamboo decorations that indicated a noble wedding when I arrived in September. It was a good example of how rituals that reproduced social hierarchy had been revived in the thirty years since the rebellion, even among the most devoutly modernist Muslims.

As we saw in chapter 7, Muhammad Idris served under Kahar Muzakkar from 1952 to 1960. After the end of the rebellion, Idris also found a position as an elementary school teacher. Like Hakim, Idris's lack of access to formal education credentials in his youth has meant that his career as a schoolteacher could not go beyond the elementary level. Like Hakim as well, he found personal fulfillment through artistic expression. He played the traditional Makassar flute and coached the village youth and maidens in traditional dances. In the 1980s, he served as the principal adviser to a film crew that came from Makassar to film the *salonreng*, the traditional wedding dance of Ara.

The examples of Muhammad Idris and Abdul Hakim show how a youthful commitment to an explicit ideological program tends to be moderated as one ages and absorbs an ever more complex under-standing of the implicit symbolic system of a culture. From the strict enforcement of a simple and universal code of religious law, they shifted their interests toward the preservation and creation of traditional aesthetic forms. The militants who helped pull down the walls surrounding the tombs of saints in the 1950s and who strictly forbade the veneration of saints as *shirk*, idolatry, now saw such behavior as the product of tolerable ignorance on the part of the "people."

Epilog

As it turned out, Habibie's Golkar proved to be only a temporary vehicle for these sentiments. Jusuf Kalla survived the scandal of 2000. In the presidential elections of 2004, he was selected by General Susilo Bambang Yudhoyono to run as his vice presidential nominee on the Partai Demokrat ticket. In the first round of balloting on July 5, their ticket came in first nationally, with 34 percent of the vote. Megawati's PDI-P ticket came in second at 27 percent of the vote, General Wiranto's Golkar came in third at 22 percent, Amien Rais's

PAN came in fourth at 15 percent, and Hamzah Haz's PPP came in fifth at 3 percent.

In Ara, the vote for Yudhoyono and Kalla in the July 2004 round of balloting was an overwhelming 84 percent, while Wiranto of Golkar finished a distant second with just 7 percent. During the run-off between Megawati and Yudhoyono on September 20, 2004, Yudhoyono won over 90 percent of the vote in the *kecamatan* of Bonto Bahari and in the *kabupaten* of Bulukumba. I interpret this result as a vote for a balanced relationship between Yudhoyono's commitment to maintaining central authority and Kalla's commitment to a measure of provincial autonomy and to a modernist form of Islam.

Chapter 9

Conclusion: Narrative, Ritual, and Models of the Self

Symbolic knowledge among the Makassar draws on many different types of experience. The oldest and most localized types of symbolic knowledge are the traditional myths and rituals that are based on people's practical knowledge of the immediate natural environment and of traditional techniques for exploiting it. Localized kin groups and regional polities reproduced themselves through these myths and rituals. When gunpowder technology enabled the regional polity of Gowa to grow into an empire, its rulers decided to join the symbolic world defined by the sacred scriptures of Islam. The internal development of this symbolic world was governed according to a completely different set of rules from the symbolic world of tropical Austronesia.

The Islamic rulers of South Sulawesi had to contend with the predatory designs of the Dutch East India Company from the moment of their conversion. The power of the VOC was due in large part to its single-minded pursuit of profits, its systematic methods of collecting and dispensing factual information, and its ability to efficiently allocate scarce resources among competing economic, political, and military needs. For 200 years, Makassar rulers continued to resist the bureaucratic power of the VOC through the traditional means of forming strategic marital alliances and the charismatic means of uniting diverse groups under the banner of Islam.

These tactics came to an end with the British occupation of 1811–1816. During the nineteenth century, most local elites made their peace with the Dutch colonial government and acquiesced in the distinction it drew between secular political matters and private religious matters. What resistance there was to European rule shifted from the royal courts to popular Islamic movements such as the Sammaniyya.

When European methods of schooling and military training were made available to Indonesians in the twentieth century, however, the colonial distinction between politics and religion was challenged by Islamic nationalists. They questioned the legitimacy of the entire colonial enterprise and sought to enshrine *shariah* law in the constitution of the state. But they accepted the basic principle of global nationalism, that all individuals owed their chief loyalty to their nation, and not to their local community or to the world Islamic community. At the dawn of the twenty-first century, the accelerating flow of goods and information produced by the Internet were leading to new forms of global ideology such as extremist Islamism.

When presented as a linear narrative like this, it is easy to forget that local, regional, cosmopolitan, and global forms of symbolic knowledge do not succeed one another in time, but coexist through time. All four levels of experience are a part of contemporary life in Indonesia, as they are everywhere else in the world. Each generates a distinct set of experiences that are interpreted according to symbolic schemes that change at different rates through time. Makassar political elites have been able to draw on traditional models of Indic kingship since the eleventh century, on charismatic models of Islamic authority since sixteenth century, and on the rational-bureaucratic model of the Dutch colonial state since the nineteenth century. Traditional, charismatic, and bureaucratic forms of authority thus form an array of alternative methods for claiming legitimacy, not stages in an evolutionary process.

At a finer level of detail, it is also clear that Austronesian, Indic, Islamic, and bureaucratic models of the state each exist in multiple forms. They provided the raw materials out of which competing political elites create the dominant ideology at each moment of time. The multiplicity of the models available to political actors greatly expands their ability to improvise solutions to novel problems using an existing set of symbolic tools (Lévi-Strauss 1962; compare Lambek 1993). Political leaders are neither engineers who sit down and consciously design the symbolic tools they need from scratch, nor are they actors who read from an existing cultural script. To be successful, they must not only be able to understand and manipulate material resources and the rational decisions of other actors, but they must also be deeply attuned to and able to manipulate the unconscious symbolic archive of their culture.

In ordinary times, one or another of the many available symbolic models of legitimate political authority may enjoy a position of hegemony. But those who refuse to accept this hegemonic model have available to them a range of alternative models with which to formulate their resistance. Political actors thus do not resist hegemonic models in the name of some simple, practical kind of everyday knowledge. They resist them in the name of an alternative moral order. In times of political and economic upheaval, actors may take the revolutionary step of attempting to replace one hegemonic model with another. This is what happened in Ara in 1954, when Islamic militants attempted to impose their modernist Islamic model on society as a whole, provoking the followers of the Amma Lolo to counter them in the name of their own version of an ancient Austronesian model instead. As more peaceful conditions returned in the 1960s, each of these models lapsed back into a marginalized position as Suharto's peculiar mix of traditionalism and developmentalism was imposed from above. When political repression lifted in the late 1990s, local political actors were again able to advocate openly for any one of a number of alternative ideal models that had been available to them all the time.

Distinguishing conscious ideological manipulation from unconscious symbolic knowledge and practice allows one to escape the false opposition between Barth's "transactional" view of politics, in which rational actors manipulate culture for self-evident ends such as wealth and power, and Geertz's dramaturgical model of traditional states as theaters in which actors perform roles written for them by their culture (Barth 1960; Geertz 1980; see Errington 1989 for an application of Geertz's model to South Sulawesi). On the one hand, it is clear that actors do attempt to manipulate culture, and that some actors are much better at this than others. On the other hand, it is equally clear that the most effective political actors are the ones who are able to bring these unconscious symbolic models to conscious awareness, and to manipulate them along with a variety of other practical factors such as military force and economic resources to achieve their political objectives. In fact, there is no need to choose between these two positions. Even when political leaders engage in wholly cynical manipulation of a population's symbolic models, such manipulation is unlikely to be efficacious unless the leaders understand the models they are using from the inside, "intuitively."

Ritual Experience and Models of the Self

The intuitive understanding of the symbolic complexes discussed in this book is acquired through immediate ritual experience. It is through such experiences that narratives about individuals who lived in remote times and places become concrete models for different ways of conceptualizing the self. In the course of this book, we have encountered many individuals who shifted their commitment from an ideal model rooted in one symbolic complex to an ideal model rooted in another over the course of their lifetime. As they shifted from one frame of symbolic reference to another, their conception of their essential identity and of their ultimate goals also shifted. While a detailed exploration of these issues must await a future book, I want to briefly summarize the way I see the narratives discussed in this book relating to different modes of ritual experience and to competing models of the self (see Gibson 1994, 1995).

The sacred scriptures of Islam include both the divinely inspired messages contained in the Koran and the *hadith* and *sunna*, little narratives about the words and actions of the Prophet that serve as a supplementary guide to righteous conduct. Five public ritual acts are enjoined on all believers by the Koran: the confession of the faith, the five daily prayers, the annual observance of the month of fasting, the payment of alms, and the pilgrimage to Mecca. The *hadith* provide a guide for how to conduct innumerable details of daily life in accordance with the practice of the Prophet, including hygiene, diet, and etiquette. For those who are able to perform the rites of the *hajj*, the entire Koranic text comes alive in a new way, since the worshippers are able to tread the same ground and see the same sites as the Prophet. The Koran is filled with eschatological passages reminding sinners of the torments of the grave and of the pleasures and torments that will be meted out on Judgment Day. When read out during funeral rituals, these passages are meant to evoke a visceral fear of dying in a state of sin, and to encourage people to continually step back from the passions of daily life and look to the implications of their actions for their place in the afterlife. Religious piety splits the self into sacred and profane parts that are not always easy to reconcile. As Weber realized, the very existence of a world religious existence creates numerous occasions where the conflict between immediate and ultimate ends cause the self to become reflective and to engage in deliberate choice between courses of action that are evaluated differently according to local and world norms.

In conclusion, I would like to outline some of the implications the rituals associated with each of the ideal models outlined in this book have for the formation of multiple selves. First, the three *shaikhs* from Sumatra who are credited with the conversion of South Sulawesi to Islam brought with them the doctrines of Hamzah Fansuri and of the Mughal Emperor Akbar. These doctrines enabled the traditional rulers of South Sulawesi to act as the charismatic heirs of the Prophet Muhammad and of the Perfect Men who succeeded him as the mystical axes around which creation revolved. Tradition holds that they achieved their first success with the ruler of Luwu', whose royal house enjoyed the highest rank in the area. Sultan Abdullah of Tallo' was next, and he proved instrumental in the conversion of all the other rulers of South Sulawesi to Islam. He did so both through the threat of force and by providing a model for how hereditary rulers could claim the charismatic authority of Islam with abandoning the traditional authority they derived from the royal origin myths and local political rituals. Following his lead, local rulers continued to be installed on the sacred rock where the founding royal ancestors had first descended from the Upperworld, but they did so while swearing an oath on the Koran.

Traditional rituals of homage to the ruler and to the royal ancestors were preserved alongside charismatic rituals of homage to local and regional *shaikhs*. The annual pilgrimage to the summit of Mount Bawakaraeng stood at the apex of this syncretic system. This site had served for centuries as the symbolic center of the Makassar people. In the fifteenth century, it became the center of the rituals dedicated to Karaeng Lowe, the Makassar version of Lord Shiva who was worshipped all around the Java Sea (Gibson 2005: 122–125). This cult still has many followers, known as the To Onto, who live on the upper slopes of the mountain. In the seventeenth century, Mount Bawakaraeng was reinterpreted as both a local manifestation of the Kaba in Mecca and as the local abode of Abd al-Qadir Jilani, the twelfth-century founder of one of the first Sufi *tariqa*. It thus became a site where Muslims could fulfill both the obligatory ritual practice of the *hajj* and pursue the optional mystical practices of the Qadiriyya.

The net result of this model of conversion was to maintain the royal court as the center of the social, political, and religious hierarchy. It allowed noble houses everywhere to continue to practice the hierarchical life-cycle rituals that reproduced their privileged relationship with the royal ancestor spirits while also observing the egalitarian Islamic rituals that marked an individual's entry into the *umma* at birth and departure at death.

Second, soon after the conversion of the kings, young men from all over South Sulawesi began traveling to Mecca to acquire religious knowledge from a source that was absolutely autonomous from the local social and political hierarchy. The piety of the "neo-Sufi" *shaikhs* they encountered in the Holy Land laid particular stress on the universal norms laid out in the *shariah* law and in spiritual lineages that tran-scended all local traditions. Cosmopolitan *ulama* such as Nur al-Din al-Raniri traveled in the opposite direction and undertook the great task of translating the scriptures into regional languages such as Malay.

Others undertook the next step of translating these texts into local languages such as Makassar and Bugis. These translations eventually provided at least a few literate inhabitants of most villages in South Sulawesi with access to cosmopolitan religious knowledge that was not controlled by the royal courts.

In the 1640s, La Madarammeng of Bone challenged the hegemony of Gowa by appropriating the cosmopolitan doctrines of the *ulama* and *shaikhs* of the Arabian Sea. While La Ma'darammeng was defeated by the armies of Gowa in the short run, the cosmopolitan model he advocated triumphed when his heir, Arung Palakka, formed an alliance with the VOC to make Bone the hegemonic power in South Sulawesi. Lineages of *kali* were founded all over the peninsula by *shaikhs* who originated in the lands surrounding the Arabian Sea, or who had at least studied there. A complementary opposition was established between the traditional authority of the *karaengs*, or hereditary rulers, and the charismatic authority of the *kali*, or religious officials.

Although village rituals surrounding birth, marriage, and death continued to be performed primarily by traditional ritual experts called *sanro*, they also all invoked both the political authority of the *karaeng* and the religious authority of the *kali*. These rituals were major occasions for substantial "customary" payments to both political and religious officials called *pangadakang* (from *adat*, custom). Local corporate groups could thus not reproduce themselves without paying tribute to both political rulers and religious authorities. The same continues to hold true for every individual, at least symbolically. Every major crisis in an individual's life is marked by rituals that require the services of a *sanro* recognized by the local community, a village official recognized by the provincial government, and an *imam* recognized by the Islamic *umma*.

Shaikh Yusuf proved that a fatherless individual from the extreme margins of the Islamic world could master the religious knowledge contained at the center and establish a new source of charismatic power that flowed through the water that emerged from his tomb and through the blood of his descendents. At the beginning of the *Riwayat Shaikh Yusuf*, the sultan of Gowa contemptuously rejects Yusuf's request to marry his daughter, Daeng Nisanga, because of his low social rank. At the end, the nobles of Gowa are all eager to drink some of the water found in his coffin, and thereby incorporate some of the charisma he had accumulated during a life of religious study and mystical practice. The body of his disciple, Tuan Rappang, also produced charismatic water that Makassar nobles were happy to ingest. The bodies of

Shaikh Yusuf and Tuan Rappang were both moved from their original places of interment and reburied at the edge of the royal cemetery of the kings of Gowa. Yusuf's royal wife, Daeng Nisanga, was buried next to him. Sultan Abd al-Jalil inaugurated the practice of making annual visits to the tomb of Shaikh Yusuf, symbolically marking the subordination of political to religious authority. But the incorporation of Yusuf's body into the royal cemetery in this way also marked the domestication of the cosmopolitan charismatic authority of the *shaikhs* by the regional traditional authority of the kings.

The branch of the Khalwatiyya founded by Shaikh Yusuf is still in existence. Its members continue to be guided through a realm of direct mystical experience by the detailed commentaries Yusuf wrote on altered states of consciousness during his exile in Sri Lanka. The Khalwatiyya Yusuf continues to be restricted to Makassar nobles of high rank. Although their numbers are few, they exercise a disproportionate, if dwindling, social influence.

Members of the general population who are excluded from membership in the Khalwatiyya Yusuf can participate in this symbolic complex by visiting the joint tombs of Shaikh Yusuf and Daeng Nisanga in order to secure their blessings. This is customarily done just after a wedding to ensure the fertility of the new union. In this way, every husband is associated with the charismatic power of the *shaikh*, every wife is associated with the traditional power of the princess, and their union is associated with the androgynous fertility of the joint tomb.

The symbolic complex of narrative and ritual practice that formed around the legacy of Shaikh Yusuf represents a synthesis of traditional and charismatic authority in which charismatic individuals were allowed to marry far above their hereditary rank, and in which the descendents of royal ancestors were prepared to acknowledge the religious superiority of charismatic *shaikhs*. Austronesian gender symbolism is reinscribed within Islamic ritual.

Third, the *Sinrili' Datu Museng* is perhaps the most complex of all the narratives analyzed in this book. I have shown how it simultaneously serves as a retelling of traditional Austronesian myths about the reunion of twins who are separated at birth; as an allegory of the charismatic Sufi path to union with the Godhead; and as an oral tradition of popular resistance to the depredations of the bureaucratic VOC. It portrays the world as divided into three mutually antagonistic spheres centered on different places: a local social hierarchy centered on a royal court that has lost its political autonomy, a cosmopolitan religious hierarchy centered on a distant holy land, and a global political hierarchy centered on the VOC headquarters in Batavia. It links the locally generated motivation of young men to marry above their station to the religiously generated motivation of young men to acquire esoteric Islamic knowledge (*ilmu*) in Mecca, and the cosmopolitan knowledge so acquired to the ability to resist the military power of European colonialism.

The conviction that esoteric knowledge can overcome raw military force is rooted in a number of concrete ritual techniques. Preindustrial techniques of warfare rely on mastering the fear of death in hand-to-hand combat. The spiritual austerities (*tapa*) taught by the Sufi orders induce otherworldly state of consciousnesses that renders the mystic relatively indifferent to life in this world. Training in the martial arts (*sila'*) induces a related state of heightened awareness that enables a fighter to parry blows with uncanny speed and accuracy. Finally, combat itself induces an altered state of consciousness. Those who survive it unscathed often attribute both their transformed psychic state and their survival to the charisma of the talismans (*jima'*) they carried into battle.

The significance of the *Sinrili' Datu Museng* goes beyond the way it weaves together gender symbolism, Islamic mysticism, and armed resistance to colonialism. As I will show in my book on Makassar life-cycle rituals, it also serves as an elaborate commentary on noble wedding rituals. These rituals are the most expensive and elaborate of all Makassar rituals, for they concern the means by which noble houses reproduce themselves. They portray each marriage as simultaneously fulfilling four different models of marriage. First, Makassar origin myths and rituals portray noble weddings as *political alliances* between houses that are formally negotiated by male authorities during elaborate rituals of betrothal. Second, they portray noble weddings as the *reunification of cousins* who were predestined for one another while still in the womb. Third, they portray noble weddings as *heroic tests* in which a man of humble origins must prove his personal courage and virility to the parents of a princess. Fourth, they portray the physical consummation of the marriage as signifying a deeper *mystical union* between the male and female principle, a union that provides a foretaste of heavenly bliss. All four of these models can be found in the story of Datu Museng and Maipa Deapati, making it entirely appropriate that it is recited during the climactic nights of a noble wedding. The *Sinrili'* provides a symbolic structure for the heightened emotions that are associated with weddings. It enables all the young noble men and women who are undergoing these emotions for the first time to interpret their personal experiences in relation to the whole history of the Makassar people.

The tombs of Datu Museng and Maipa Deapati are situated at what was once the edge of the Dutch settlement around Fort Rotterdam, and at the boundary between the land and the sea. As in the case of Shaikh Yusuf and Daeng Nisanga, the joint tombs of Datu Museng and Maipa Deapati remain as concrete residues that were created when the cosmopolitan peregrinations of the men brought them home to reunite with their royal wives in death. Their charismatic transcendence of local social and political hierarchies in life enables their remains to serve as channels for divine blessings in death.

Fourth, Sultan Ahmad al-Salih of Bone combined the traditional authority he derived from his descent from the royal houses of Bone and Gowa with the charismatic authority he derived from his patronage of a circle of cosmopolitan scholars who translated Islamic works from Arabic and Malay into Bugis and from his initiation into the Sammaniyya. While mystical knowledge continued to be restricted to the high nobility during his reign, these translations and the Sammaniyya facilitated the popularization of Islamic learning during the nineteenth century. As the colonial wars against Bone undermined the ability of the royal court to serve as a center for Islamic learning, the *shaikhs* of the Sammaniyya turned it into an institution for the transmission of religious knowledge that was independent of the state. The spread of popular mysticism was thus correlated with the growing power of the secular colonial state. It was only in light of this newly developed peaceful coexistence between a privatized sphere of religion and a secularized state that the seventeenth-century alliance between Arung Palakka and the VOC could be reinterpreted as legitimate. Arung Palakka could then be portrayed in the *Sinrili' Tallumbatua* as a pious wanderer not unlike Shaikh Yusuf, and he too could be buried near the royal cemetery of Gowa alongside his royal Gowanese wife, Daeng Talele.

During the nineteenth century, the continuing vitality of Islam in South Sulawesi rested on the "customary payments" made to local *kalis* and *imams* in return for their services during calendrical and life-cycle rituals. A central part of the services they provided was the recitation of appropriate passages from Makassar translations of classical Islamic works by al-Ghazzali and al-Raniri; of the loud, collective *dhikr* of the Sammaniyya; and, above all, of the *Maulid al-Nabi* by al-Barzanji. These scriptures encouraged every individual to measure his or her life against that of the Prophet, and to contemplate their ultimate fate on Judgment Day. Because of the way they were linked to emotionally charged experiences such as birth, marriage, death, and collective chanting sessions, the explicit religious doctrines they contained acquired a deeper inner resonance. Local social hierarchies and regional political authorities played no role in these religious scriptures. As long as the local religious leaders were left alone, they tended to adopt a quietist attitude toward social and political issues.

Fifth, the peaceful coexistence of separate social, political, and religious hierarchies was destroyed in the late nineteenth century. The colonial state made a tactical political alliance with ambitious men of low social status to undermine the traditional authority of local political rulers. The state sought to implement a new form of rational-bureaucratic authority at the local level by appointing salaried agents who could implement policies that aimed at improving the roads, controlling epidemics, and enforcing the law. The local leaders who put themselves forward as capable of doing so claimed a new form of charismatic authority that was at odds with the traditional cult of the royal ancestor cults. Haji Gama accepted his role as a dependent agent of the colonial bureaucracy, but he also claimed an autonomous kind of charismatic authority that derived from his mastery of mystical techniques (*tapa*); from his administration of the *shariah* law; from his patronage of the cult of the village *shaikh*, Bakka' Tera'; and from his performance of the *hajj* at the end of his life.

Haji Gama's fierce critique of the royal ancestor cults helped to solidify a breach between male and female forms of religious practice in Ara. In the late nineteenth century, the *karihatang*, or royal spirit medium, appears to have been a man. Ritual séances were sponsored by the village chief, and equal numbers of male and female performers participated in the processions to the tomb of Karaeng Mamampang. A woman, To Ebang, took over as *karihatang* in 1910. After Haji Gama began his campaign against the cult in the 1930s, To Ebang found it increasingly difficult to recruit the male service she needed to carry out her séances. Her granddaughter, Titi Daeng Toje', took over as *karihatang* in 1962. By the 1980s, her séances were attended almost exclusively by women. At least in Ara, the traditional authority that derived from the old myths and rituals of kingship had become associated with a dwindling group of noble women, while the charismatic authority that derived from Islamic mysticism had become the preserve of men.

Haji Gama's political and religious prestige grew so great during the 1920s and 1930s that his descendents were able to marry into the royal house of Tanaberu, and did so to acquire a measure of traditional authority. Two of his grandsons, Drs. Basri and Andi' Azikin, went on to acquire a large measure of bureaucratic authority by obtaining advanced credentials in the state school system, followed by appointment to high office in the provincial government. The traditional and bureaucratic authority acquired by this branch of the family began to overshadow the charismatic authority of Haji Gama. By 2000, Drs. Basri was trying to rewrite the epitaph of his great grandfather, Panre Abeng. The revised inscription portrayed him as a noble from the most prestigious of the traditional kingdoms, Luwu', and changed his title from *panre*, expert, to *tenri*, making him the namesake of a minister in the national government. Were Drs. Basri to succeed in this reinscription of Panre Abeng's origins, the family tomb would become a permanent marker of his noble rank and of his right to high office.

Panre Abeng's grandson by way of his daughter, Sadaria, had a different perspective. Muhammad Idris spent his youth fighting with Kahar Muzakkar to establish an Islamic state and so failed to obtain the formal credentials he needed to advance within the educational bureaucracy. In 2000, Muhammad Idris continued to insist on the purely charismatic sources of Panre Abeng's authority. On the basis of his own miraculous survival during nine years of guerilla warfare, he argued that the esoteric knowledge he acquired from his grandfather's book of incantations must have been due to the fact that this book had been passed down within the charismatic lineage founded by Haji Ahmad, the Bugis, in the seventeenth century. He was only interested in using the relationship between his ancestors and the cosmopolitan world of Islamic knowledge to explain the heightened state of awareness he had experienced in actual combat. He had no interest in the regional or even national relationships that might be claimed for his ancestors.

Sixth, the complementarity of traditional social hierarchy, colonial bureaucracy, and the cosmopolitan Islam of the late nineteenth-century *hajjis* broke down in the 1920s as formal schooling began to create a cadre of Indonesians who were exposed to the global ideologies of nationalism, socialism, and Islamic modernism. In the face of rising demands for democratic self-governance and the collapse of the market in tropical exports during the great depression, the colonial state began lose its claim to rational-bureaucratic authority. It was forced to resort to increasing levels of political repression to maintain control over the Netherlands East Indies. It reversed its hostile attitude toward the rituals and beliefs that maintained the traditional authority of local royal houses. Government functionaries were instructed to collect, preserve, codify, and implement authentic rituals to select and install hereditary rulers.

Islamic modernists continued to critique these traditional rituals, but they also expanded their critique to many Islamic practices they regarded as "innovations," meaning that they had no foundation in the core scriptures of Islam. Foremost among these were the practices of *siara*, visiting the tombs of *shaikhs* to obtain blessings; *berdiri Maulid*, standing during the recitation of texts in praise of the

Prophet to show respect for the presence of his spirit; and collective *dhikr*, chanting phrases in praise of God under the leadership of a spiritual master. These practices all smacked of *shirk*, treating human creatures like gods. During the 1930s, modernists called for the abolition of these rituals, and of the *pangadakang*, the customary payments villagers made to hereditary political and religious leaders at every major life-cycle ritual. This call was strongly resisted by both traditional and colonial authorities, for it threatened to destabilize the complementary relationships between the social, religious, and political hierarchies that had developed during the nineteenth century.

During the early 1940s, many educated Indonesians received further military training from the Japanese, which introduced them to new models of nationalism and bureaucratic statism. They refused to accept the return of any form of Dutch authority after the war was over. In South Sulawesi, many of the Islamic modernists who fought against the return of the secular colonial state in the late 1940s continued to fight against the imposition of a secular nation-state in the 1950s. They believed that the *shariah* law could provide a unified set of principles to govern conduct in all three spheres of life, if only it were correctly interpreted and applied. This meant that one had to abandon the principle of *taklid*, the unquestioning acceptance of traditional interpretations, and one had to apply individual reasoning anew to the core Islamic scriptures. In the view of the Islamic revolutionaries of this generation, there would be no need to maintain the distinction between the spheres of social hierarchy, religion, and the state once one eliminated the feudal rituals that reproduced the social hierarchy, the polytheistic practices that had crept into traditional Islam, and the secular principles that the Dutch had introduced into the legal system.

Many educated youths such as Muhammad Idris and Abdul Hakim were committed to the creation of a unified social, political, and religious order in their youth, but later came to accept the ineluctable pluralism and complexity of human experience and symbolic knowledge. They were especially sensitive to the differences between religious, scientific, and aesthetic forms of knowledge. They were able to read and reflect on Islamic scriptures and commentaries in the original Arabic and in Malay and Makassar translations. They were accomplished schoolteachers who were able to comprehend mathematical and scientific texts and teach them to their students. Finally, they had both mastered traditional Makassar art forms such as music, dance, and painting.

Despite their opposition to traditional Sufi brotherhoods, most modernists did not abandon the pursuit of the esoteric knowledge that came from mystical experience. They did eliminate all public, collective forms of mystical practice. For them, mysticism became an interiorized private quest based on silent prayer, meditation, and spiritual discipline. These practices often yielded mystical visions that allowed an individual to obtain *ilmu* from beings such as the spirits of *shaikhs, jinn, malaikat* (angels), or directly from God, without having to rely on imperfect human masters. During the 1980s, *ilmu* obtained in this way enabled a number of highly educated and highly placed civil servants that I knew to treat a range of physical and spiritual diseases.

Seventh, during the 1970s and 1980s, Suharto consolidated his personal control over the civil, police, and military bureaucracies. He also tried to acquire a combination of traditional, charismatic, and bureaucratic authority by combining elements from the Austronesian, Indic, Islamic, and Dutch pasts to create a uniform national culture. At the end of the 1980s, villagers in South Sulawesi were still rather skeptical of Suharto's religious piety, which privileged traditional Javanese mysticism over cosmopolitan Islam. Villagers were very aware of the fact that the Cold War was in its final throes, and that the *mujahidin* of Afghanistan had played a key role in the demise of atheistic socialism.

The Cold War was the first truly global war since it spread to every corner of Africa, Asia, and Latin America. Its end was marked by the demise of Marxism as a global ideology of resistance to Euro-American capitalism, and by the emergence of a globalized Islamist ideology that preached resistance to neocolonialism. The foundations of this movement were laid during the 1970s and 1980s as rising oil revenues enabled the Saudi regime to subsidize the teaching of Wahhabi doctrines in mosques and *madrassas* around the world. Saudi prestige had grown so great by the 1990s that Suharto himself sought the blessings of King Fahd and formed an alliance with Islamist generals sympathetic to Wahhabi teachings.

In 2000, my friends in Ara seemed persuaded of the sincerity of Suharto's "Islamic turn." They had supported his chosen successor, Habibie, in the national elections of 1999. They were unperturbed, however, when the party Suharto had created, Golkar, faded into oblivion in the 2000s and happily switched their support to General Susilo Bambang Yudhoyono, in part because his running mate, Jusuf Kalla, was a favorite son from South Sulawesi and had a solid reputation as an Islamic modernist.

Symbolic Knowledge and Individual Agency

The historical and symbolic materials discussed in this book raise a number of questions about the role of narratives in the formation of different kinds of subjects. The Makassar self is just as much the product of a long history of ritual and myth as is the state and there are as many competing models of the ideal self as there are of the ideal state. These models specify not only how the external world of social and political relations should be ordered, but also how the internal world of subjective experience and aspirations should be organized. They specify whether one should be an industrious craftsman who seeks prestige and religious merit by sponsoring expensive rituals; a link in a hierarchical chain of local ancestors or cosmopolitan saints who maintains social relations with the spirits of the dead through rituals of veneration; a humble seeker who pursues mystical knowledge in this life through the practice of austerities; or a well-educated autonomous individual competing for a place in the state bureaucracy and in the afterlife by following a set of rational and explicit rules.

The narratives also raise questions about what happens to a person's sense of self when he or she shifts allegiance from one ideal model to another. For example, academic discipline forms a certain kind of competitive moral agent. An ambitious youth might devote himself to academic achievement in the school system in order to achieve high political office. Later in life, the same individual might decide that the rewards of a bureaucratic career are insufficient and develop an interest in mystical experience. But this requires the individual to cultivate a very different set of dispositions and aspirations. These might lead him to question the value of his life in the bureaucracy at an even deeper level, to the point of abandoning it altogether. In this case, the individual might be said to have undergone a religious conversion, an inner transformation that is so fundamental that he has become a "new man." Conversely, an individual may become increasingly cynical with age and devote himself to the manipulation of symbolic knowledge to serve his practical ends.

Whether individuals tend toward a more spiritual or a more political practice, their older orientations are seldom left entirely behind. As people are exposed to and adopt different ritual, disciplinary, and symbolic practices, their inner life tends to become more complex. The complexity of local cultures and selves that are exposed to global circuits of knowledge and power continually increases over time as new symbolic models are introduced without displacing the old ones. As the turbulent political and religious history of South Sulawesi demonstrates so clearly, the relationship between politics and religion is contested in every generation.

References

Abbreviations

BKI Bijdragen van het Koninklijk Instituut voor Taal-, Land- en Volkenkunde.
KITLV Koninklijk Instituut voor Taal-, Land- en Volkenkunde.
TNI Tijdschrift voor Taal-, Land- en Volkenkunde van Nederlandsche Indie.
VKIT Verhandelingen van het Koninklijk Instituut voor Taal-, Land- en Volkenkunde.

* * *

Abdurrazak Daeng Patunru. [1960]. *Sedjarah Gowa*. Makassar: Jajasan Kebudayaan Sulawesi Selatan dan Tenggara.
Abu Aufa ash-Shiddiquie. 1986. *Maulid an-Nabi oleh Jaffar al-Barzanji*. Kudus, Indonesia.
Abu Hamid. 1983. "Sistem Pendidikan Madrasah dan Pesantren di Sulawesi Selatan." In Taufik Abdullah, ed. *Agama dan Perubahan Sosial*. Jakarta: Rajawali.
———. 1994. *Syekh Yusuf: Seorang Ulama, Sufi dan Pejuang*. Jakarta: Yayasan Obor Indonesia.
Abu-Lughod, Janet. 1989. *Before European Hegemony: The World System A.D. 1250–1350*. New York: Oxford University Press.
Alfian. 1969. "Islamic Modernism in Indonesian Politics." Ph.D. Thesis, University of Wisconsin, Madison.
Alter, Joseph. 2004. *Yoga in Modern India: The Body between Science and Philosophy*. Princeton: Princeton University Press.
Andaya, Leonard. 1979. "A Village Perception of Arung Palakka and the Makassar War of 1666–1669." In Anthony Reid and David Marr, eds. *Perceptions of the Past in Southeast Asia*. Singapore: Heinemann Education Books, 360–378.
———. 1981. *The Heritage of Arung Palakka: A History of South Sulawesi in the Seventeenth Century*. VKIT 91. The Hague: Martinus Nijhoff.
Anderson, Benedict. 1985. "Further Adventures of Charisma." In his *Language and Power*. Ithaca: Cornell University Press.
———. 1991. *Imagined Communities*. Revised edition. London: Verso.

Andi Shadiq Kawu. 1989. "Persepsi Islam oleh Amma Towa." *Mimbar Karya* June 25, 1989. Ujung Pandang, Sulawesi Selatan.

Arberry, Arthur. 1957. *Revelation and Reason in Islam*. London: George Allen and Unwin.

Arief, Aburaerah, ed. 1981. *Kisah Syekh Mardan*. Jakarta: Departemen Pendidikan dan Kebudayaan.

Asad, Talal. 1993. Genealogies of Religion: Discipline and Reasons of Power in *Christianity and Islam*. Baltimore: Johns Hopkins University Press.

al-Attas, Syed Muhammad Nauguib. 1963. "Two famous orders in Malaya." In his *Some Aspects of Sufism*. Singapore: Malaysian Sociological Research Institute.

―――. 1966. *Raniri and the Wujudiyyah of 17th Century Aceh*. Singapore: Monographs of the MBRAS III.

―――. 1970. *The Mysticism of Hamsah Fansuri*. Kuala Lumpur: University of Malaya Press.

―――. 1986. *A Commentary on the Hujjat al-Siddiq of Nur al-Din al-Raniri*. Kuala Lumpur: Ministry of Culture.

Azra, Azyumardi. 1992. "The Transmission of Islamic Reformism to Indonesia." Ph.D. Thesis, Columbia University, New York.

―――. 2004. *The Origins of Islamic Reformism in Southeast Asia*. Honolulu: University of Hawai'i Press.

Bakkers, J.A. 1866. "Het Leenvorstendom Boni." *Tijdschrift voor Indische Taal-, Land- en Volkenkunde* 15: 1–208.

Barth, Frederick. 1960. *Political Leadership among Swat Pathans*. London: Athlone Press.

Batten, R.J.C. 1938. "Aanvullende memorie van overgave van de onderafdeeling Boeloekoemba 1938." Unpublished manuscript held in the Koninklijke Institute voor de Tropen, Amsterdam.

Berger, Peter. 1967. *The Sacred Canopy: Elements of a Sociological Theory of Religion*. Garden City, NY: Doubleday.

Berigten. 1855. "Berigten Omtrent den Zeeroof in den Nedelandsch-Indischen Archipel over 1852 en 1853." *Tijdschrift voor Indische Taal-, Land- en Volkenkunde* 3: 1–31.

Bloch, Maurice. 1971. *Placing the Dead: Tombs, Ancestral Villages and Kinship Organization in Madagascar*. London: Seminar Press.

―――. 1986. *From Blessing to Violence: History and Ideology in the Circumcision Ritual of the Merina of Madagascar*. Cambridge: Cambridge University Press.

―――. 1998. *How We Think They Think: Anthropological Approaches to Cognition, Memory, and Literacy*. Boulder: Westview Press.

Blok, Roelof. [1759] 1817. *History of the Island of Celebes*. Trans. by J. von Stubenvoll in 4 vol. Calcutta: Calcutta Gazette Press.

―――. [1759] 1848. "Beknopte Geschiedenis van het Makassaarsche Celebes en Onderhoorigheden." *TNI* 10: 3–77.

Bowen, John. 1987. "Islamic Transformations: From Sufi Poetry to Gayo Ritual." In Rita Smith Kipp and Susan Rodgers, eds. *Indonesian Religions in Transition*. Tucson: University of Arizona Press, 113–135.

————. 1993. *Muslims through Discourse: Religion and Ritual in Gayo Society.* Princeton: Princeton University Press.

Braginsky, Vladimir. 1990. "Hikayat Shah Mardan as a Sufi Allegory." *Archipel* 40: 107–136.

————. 1999. "Toward the Biography of Hamzah Fansuri." *Archipel* 57: 135–175.

van den Brink, H. 1943. *Dr. Benjamin Frederik Matthes.* Amsterdam: Nederlandsch Bijbelgenootschap.

van Bruinessen, Martin. 1987. "Bukankah orang Kurdi yang Mengislamkan Indonesia?" *Pesantren* 4(4): 43–53.

————. 1991. "The Tariqa Khalwatiyya in South Celebes." In Harry Poeze and Pim Schoorl, eds. *Excursies in Celebes: Een Bundel Bijdragen bij het Afscheid van J. Noorduyn als Directeur-Secretaris van het KITLV.* VKIT 147. Leiden: KITLV Press.

————. 1995. "*Shari'a* court, *tarekat* and *pesantren*: Religious Institutions in the Banten Sultanate." *Archipel* 50: 165–199.

Brunsveld van Hulten, P., ed. 1853. "Extract uit de Generale Resolutien des Casteels Batavia in Rade van India op Vrigdag den 23 Maart Ao. 1759"; "Instructie voor den Onderkoopman Willem Delfhout"; and "Inlandsche Wetten bij de Hoven van Bonie en Goa van Aloude Tijden in Gebruik, tot Huidigen Dagen." *Het Recht in Nederlandsch Indie* 4(8): 83–119.

Buddingh, S.A. 1843. "Geschiedenis: Het Nedeerlandsch Gouvernement van Makassar op het Eiland Celebes." *TNI* 5: 411–458, 531–536, 663–689.

Bulbeck, Francis David. 1990. "The Landscape of the Makassar War." *Canberra Anthropology* 13(1): 78–99.

————. 1992. "A Tale of Two Kingdoms: The Historical Archeology of Gowa and Tallok, South Sulawesi, Indonesia." Ph.D. Thesis, Australian National University, Canberra.

Buyers, C. 2000–2005. *The Royal Ark Royal and Ruling Houses of Africa, Asia, Oceania and the Americas.* [Including Web sites on the royal genealogies of Bone, Gowa, Tallo' and Bima.] http://www.4dw.net/royalark/; accessed August 1, 2006.

Carsten, Janet. 1997. *The Heat of the Hearth.* Oxford: Oxford University Press.

Casanova, José. 1994. *Public Religions in the Modern World.* Chicago: University of Chicago Press.

Cense, A.A. 1950. "De Verering van Sjaich Jusuf in Zuid Celebes." In Samuel van Ronkel, ed. *Bingkisan Budi.* Leiden: A.W. Sijtohoff, 50–57.

Chabot, Hendrik. 1950. *Verwantschap, Stand en Sexe in Zuid-Celebes.* Groningen: J.B. Wolters.

————. 1996. *Kinship, Status and Gender in South Celebes.* Leiden: KITLV Press.

Chaudhuri, K.N. 1985. *Trade and Civilization in the Indian Ocean.* Cambridge: Cambridge University Press.

Chomsky, Noam. 1993. *Year 501: The Conquest Continues.* Boston: South End Press.

van Clootwijk, Jan. [1755] 1919. "Companiesverzameling uit 1755 en Bijbehoorende Instructie." In *Adatrechtbundel* XVII, Series P, Zuid-Celebes No. 11: 150–151.

Collins, George. 1936. *East Monsoon*. London: Jonathan Cape.

———. 1937. *Makassar Sailing*. London: Jonathan Cape.

Crone, Patricia and Martin Hinds. 1986. *God's Caliph: Religious Authority in the First Centuries of Islam*. Cambridge: Cambridge University Press.

Cummings, William. 2001. "The Dynamics of Resistance and Emulation in Makassar History." *Journal of Southeast Asian Studies* 32(2): 423–435.

Cunnningham, Clark. 2000. "Foreword." In Rosemary Joyce and Susan Gillespie, eds. *Beyond Kinship: Social and Material Reproduction in House Societies*. Philadelphia: University of Pennsylvania Press, vii–ix.

Digby, Simon. 1986. "The Sufi Shaikh as a Source of Authority in Mediaeval India." In Marc Gaborieau, ed. *Islam and Society in South Asia*. Editions de EHESS: Paris, 57–77.

van Dijk, Kees. 2001. *A Country in Despair: Indonesia Between 1997 and 2000*. Leiden: KITLV Press.

Donselaar, W.M. 1855. "Beknopte Beschrijving van Bonthain en Boelecomba op Zuid Celebes." *BKI* 3: 163–187.

Douglas, Mary. 1973. "Self-Evidence." *Proceedings of the Royal Anthropological Institute for 1972*. New York: Vintage, 27–43.

Drewes, G.W.J. 1926. "Sech Joesoep Makassar." *Djawa* 6: 83–88.

———. 1992. "A Note on Muhammad al-Samman, His Writings, and the 19th Century Sammaniyya Practices, Chiefly in Batavia, According to Written Data." *Archipel* 43: 73–88.

Dumont, Louis. 1972. Homo Hierarchicus: The Caste System and Its Implications. London: Paladin.

———. 1975. "Preface to *The Nuer*." In J.H.M. Beattie and R.G. Lienhardt, eds. *Studies in Social Anthropology: Essays in Memory of E.E.Evans-Pritchard*. Oxford: Clarendon Press, 328–342.

Durkheim, Émile. [1915] 1995. *The Elementary Forms of the Religious Life*. Trans. Karen Fields. New York: The Free Press.

Eaton, Richard. 1978. *Sufis of Bijapur 1300–1700*. Princeton: Princeton University Press.

———. 1993. *The Rise of Islam and the Bengal Frontier 1204–1760*. Berkeley: University of California Press.

Eerdmans. n.d. "Geschiedenis van Bone (met Geslachtslijst Vorsten)." Unpublished manuscript held by the KITLV, Leiden.

Elson, Robert. 2001. *Suharto: A Political Biography*. Cambridge: Cambridge University Press.

Engelhard, H.E.D. 1884a. "Mededelingen over het Eiland Saleijer." *BKI* 32: 1–510.

———. 1884b. "De Staatkundige en Economische Toestand van het Eiland Saleijer." *De Indische Gids* 6: 519–525, 817–842.

Ernst, Carl. 1992. *Eternal Garden: Mysticism, History and Politics at a South Asian Sufi Center*. Albany: SUNY Press.

Errington, Shelly. 1989. *Meaning and Power in a Southeast Asian Realm*. Princeton: Princeton University Press.

Feener, R. Michael. 1998–1999. "Shaykh Yusuf and the Appreciation of Muslim 'Saints' in Modern Indonesia." *Journal for Islamic Studies* 18–19: 112–131.

Foucault, Michel. 1977. *Discipline and Punish: The Birth of the Prison*. Penguin Books.

———. 1978. *The History of Sexuality, Volume I: An Introduction*. New York: Pantheon.

———. 1991. "Governmentality." In Graham Burchell, Colin Gordon, and Peter Miller eds. *The Foucault Effect: Studies in Governmentality*. Chicago: University of Chicago Press, 87–104.

Fowden, Garth. 1993. *Empire to Commonwealth: Consequences of Monotheism in Late Antiquity*. Princeton: Princeton University Press.

Fox, Richard. 1985. *Lions of the Punjab*. Berkeley: University of California Press.

Gade, Anna. 2004. *Perfection Makes Practice: Learning, Emotion and the Recited Qur'an in Indonesia*. Honolulu: University of Hawaii Press.

Geertz, Clifford. 1980. *Negara: The Theater State in Nineteenth-Century Bali*. Princeton: Princeton University Press.

van Gennep, Arnold. 1909. *Les Rites de Passage*. Paris: Noury.

Gervaise, Nicolas. [1688] 1701. *An Historical Description of the Kingdom of Makasar*. London: Thomas Leigh and D. Midwinter. Reprinted in 1971 by Gregg International Publishers Ltd.

Gibson, Thomas. 1989. "Are Social Wholes Seamless?" In Sherry Ortner, ed. *Author Meets Critics: Reactions to "Theory in Anthropology since the Sixties."* Working Paper of the Center for the Comparative Study of Social Transformations No. 32. Ann Arbor: University of Michigan.

———. 1994. "Childhood, Colonialism and Fieldwork among the Buid of the Philippines and the Konjo of Indonesia." In Jeannine Koubi and Josiane Massard, eds. *Enfants et sociétés d'Asie du Sud-Est*. Paris: L'Harmattan, 183–205.

———. 1995. "Having Your House and Eating It: Houses and Siblings in Ara, South Sulawesi." In Janet Carsten and Stephen Hugh-Jones, eds. *About the House: Buildings, Groups, and Categories in Holistic Perspective*. Cambridge: Cambridge University Press, 197–213.

———. 2000. "Islam and the Spirit Cults in New Order Indonesia: Global Flows vs. Local Knowledge." *Indonesia* 69: 41–70.

———. 2005. *And the Sun Pursued the Moon: Symbolic Knowledge and Traditional Authority among the Makassar*. Honolulu: University of Hawaii Press.

Gluckman, Max. 1963. *Custom and Conflict in Africa*. Oxford: Blackwell.

Goedhart, O.M. [1901] 1933. "De Pangadakang in Bira." *Adatrechtbundels* 36 Series P No. 68: 222–243.

———. [1920] 1933. "De Inlandsche Rechtsgemeenschappen in de Onderafdeeling Boeloekoemba." *Adatrechtbundels* 36 Series P No. 64: 133–154.

Goldzihir, Ignaz. 1971. "Veneration of Saints in Islam." In his *Muslim Studies, Volume II.* London: George Allen and Unwin, 245–341.

Goody, Jack. 1987. *The Interface between the Written and the Oral.* Cambridge: Cambridge University Press.

Gramsci, Antonio. 1971. *Selections from the Prison Notebooks of Antonio Gramsci.* New York: International Publishers.

Guillot, Claude, and Ludvik Kalus. 2000. "La Stèle Funéraire de Hamzah Fansuri." *Archipel 60*: 3–24.

Hamka [Haji Abdul Malik Karim Amrullah]. 1961. *Sejarah Ummat Islam.* Bukit Tinggi-Jakarta: NV Nusantara.

———. 1963. *Dari Perbendaharaan Lama.* Medan: Pertjetakan Madju.

———. 1965–1982. *Tafsir Al Azhar, Volume IX.* Jakarta: Panji Masyarakat.

Hamonic, Gilbert. 1985. "La Fête du Grand Maulid à Cikoang." *Archipel* 29(1): 175–191.

Harvey, Barbara. 1974. "Tradition, Islam and Rebellion: South Sulawesi 1950–1965." Ph.D. Thesis, Cornell University, Ithaca: New York.

Heer, Nicholas. 1979. *The Precious Pearl: al-Jami's al-Durrah al-Fakhirah.* Albany: SUNY Press.

Heersink, Christian. 1995. "The Green Gold of Selayar: A Socio-Economic History of an Indonesian Coconut Island c. 1600–1950." Ph.D. Thesis, Vrije Universiteit te Amsterdam.

Hefner, Robert. 2000. *Civil Islam: Muslims and Democratization in Indonesia.* Princeton: Princeton University Press.

Hodgson, Marshall. 1974. *The Venture of Islam.* Volume I: *The Classical Age of Islam*; Volume II: *The Expansion of Islam in the Middle Periods*; Volume III: *The Gunpowder Empires and Modern Times.* Chicago: Chicago University Press.

van Hoevell, W.R. 1854. "Bijdragen tot de Geschiedenis van Celebes." *TNI* 16(9): 149–186; (10): 213–253.

Horridge, Adrian. 1979. *The Konjo Boatbuilders and the Bugis Prahus of South Sulawesi.* Maritime Monographs and Reports No. 40. Greewich: National Maritime Museum.

Hourani, Albert. 1962. *Arabic Political Thought in the Liberal Age.* Cambridge: Cambridge University Press.

Ileto, Reynaldo. 1979. *Pasyon and Revolution: Popular Movements in the Philippines.* Quezon City: Ateneo de Manila University Press.

IJzereef, Willem. 1987. "Power and Political Structure in the Kingdom of Bone, 1860–1949." Unpublished paper presented at the Royal Institute of Linguistics and Anthropology International Workshop on Indonesian Studies No. 2.

Ito, T. 1978. "Why Did Nuruddin ar-Raniri Leave Aceh in 1054 A.H.?" *BKI* 134: 489–492.

Johns, Anthony. 1961. "Muslim Mystics and Historical Writing." In D.G.E. Hall, ed. *Historians of South East Asia.* London: Oxford University Press, 37–49.

————. 1965. *The Gift Addressed to the Spirit of the Prophet*. Canberra: The Australian National University.

————. 1975. "Islam in Southeast Asia." *Indonesia* 19: 33–55.

Jones, Russell. 1979. "Ten Conversion Myths from Indonesia." In Nehemia Levtzion, ed. *Conversion to Islam*. New York: Holmes and Meier, 129–158.

de Josselin de Jong, P.E. 1977. "Structural Anthropology in the Netherlands: Creature of Circumstance." In P.E. de Josselin de Jong, ed. *Structural Anthropology in the Netherlands*. The Hague: Martinus Nijhoff, 1–29.

de Josselin de Jong, P.E. and H.F. Vermeulen. 1989. "Cultural Anthropology at Leiden University: From Encyclopedism to Structuralism." In Willem Otterspeer, ed. *Leiden Oriental Connections 1850–1940*. Leiden: E.J. Brill, 280–316.

Joyce, Rosemary and Susan Gillespie, eds. 2000. *Beyond Kinship: Social and Material Reproduction in House Societies*. Philadelphia: University of Pennsylvania Press.

Kaestle, Carl. 1973. *Joseph Lancaster and the Monitorial School Movement: A Documentary History*. New York: Teachers College Press, Columbia University.

van Kan, J. 1935. *Uit de Rechtsgeschiedenis der Compagnie*. Tweede Bundel: Rechtsgeleerd Bedrijf in de Buitencomptoiren. Bandoeng: A.C. Nix.

Kaptein, Nico. 1992. "The *Berdiri Maulid* Issue Among Indonesian Muslims in the Period from Circa 1875 to 1930." *BKI* 148: 124–153.

Kathirithamby-Wells, J. 1970. "Ahmad Shah ibn Iskandar and the Late 17th Century 'Holy War' in Indonesia." *Journal of the Malaysian Branch of the Royal Asiatic Society* 48(1): 48–63.

Kiefer, Thomas. 1973. "*Parrang Sabbil*: Ritual Suicide among the Tausug of Jolo." *BKI* 129: 108–123.

de Klerck, E.S. 1938. *History of the Netherlands East Indies, Volume II*. Rotterdam: W.L. and J. Brusse.

Knappert, Jan. 1971. *Swahili Islamic Poetry, Volume I*. Leiden: E.J. Brill.

Kniphorst, J. 1876. "De zeeroof in den Indischen archipel." *TNI* N.S. 5(2): 355–387.

Knysh, Alexander. 1999. *Ibn Arabi in the Later Islamic Tradition*. Albany: State University of New York Press.

Laffan, Michael. 2003. *Islamic Nationhood and Colonial Indonesia: The Umma Below the Winds*. London: RoutledgeCurzon.

Lambek, Michael. 1993. *Knowledge and Practice in Mayotte: Local Discourses of Islam, Sorcery and Spirit Possession*. Toronto: University of Toronto Press.

Leach, Edmund. 1954. *Political Systems of Highland Burma*. London: Athlone Press.

Lenin, Vladimir. 1978. *What Is to Be Done?* New York: International Publishers.

Le Roux, C. 1930. "De Rijksvlaggen van Bone." *TNI* 70: 205–226.

Lévi-Strauss, Claude. 1964. *Totemism*. Trans. from the French by Rodney Needham. Harmondsworth: Penguin.

————. 1966. *The Savage Mind*. London: Weidenfeld and Nicholson.

Lévi-Strauss, Claude. 1982. "The Social Organization of the Kwakiutl." In his *The Way of the Masks*. Seattle: University of Washington Press.

———. 1983. Histoire et ethnologie. *Annales* 38(2): 1217–1231.

———. 1987. *Anthropology and Myth*. Oxford: Basil Blackwell.

Ligtvoet, A. [1875] 1987. "Bijdragen tot de Geschiedenis van het Sultanaat Sumbawa." In J. Noorduyn, ed. *Bima en Sumbawa*. VKIT 129. Dordrecht: Foris Publications.

———. 1880. "Transcriptie van het Dagboek der Vorsten van Gowa en Tello." *BKI* 28: 85–259.

Lombard, Denys. 1967. *Le Sultanat d'Atjéh au temps d'Iskandar Muda 1607–1636*. Paris: École Française d'Extrême-Orient.

Lukens-Bull, Ronald. 2005. A Peaceful Jihad: Negotiating Identity and Modernity in Muslim Java. New York: Palgrave Macmillan.

MacNeill, William. 1982. *The Pursuit of Power*. Chicago: University of Chicago Press.

Maeda, N. 1984. "Traditionality in Bugis society." In *Transformations of the Agricultural Landscape in Indonesia*. Kyoto: Center for Southeast Asian Studies, 109–122.

Makdisi, George. 1974. "Ibn Taimiya: a Sufi of the Qadiriya Order." *American Journal of Arabic Studies* 1: 118–129.

Manguin, Pierre-Yves. 1993. "The Vanishing *Jong*: Insular Southeast Asian Fleets in Trade and War (Fifteenth to Seventeenth Centuries). In Anthony Reid, ed. *Southeast Asia in the Early Modern Era*. Ithaca: Cornell University Press, 197–213.

Manyambeang, A.K. and Abdul Rahim Mone. 1979. *Lontarak Patturioloanga ri TuTalloka*. [Text and translation into Bahasa Indonesia of the Chronicle of Tallo']. Jakarta: Departemen Pendidikan dan Kebudayaan.

Martin, B.G. 1972. "A Short History of the Khalwati Order of Dervishes." In Nikki Keddie, ed. *Scholars, Saints, and Sufis: Muslim Religious Institutions in the Middle East since 1500*. Berkeley: University of California Press, 275–306.

Matthes, Benjamin. 1860. *Makassaarsch Chrestomathie*. Amsterdam: Nederlandsch Bijbelgenootschap.

———. [1883] 1943. "Inleiding op Eenige Proeven van Boegineesceh en Makasaarsche Poezie." Bijlage 23 in H. van den Brink, *Dr. Benjamin Frederik Matthes*. Amsterdam: Nederlandsch Bijbelgenootschap.

———. [1885] 1943. "Boegineesche en Makassaarsche legenden." Bilage 27 in H. van den Brink, *Dr. Benjamin Frederik Matthes*. Amsterdam: Nederlandsch Bijbelgenootschap.

Mattulada. 1976. *Agama Islam di Sulawesi Selatan*. Laporan Proyek Penelitian Peranan Ulama dan Pengajaran Agam Islam id Sulawesi Selatan. Fakultas Sastra, Universitas Hasanuddin. Typescript.

Mauss, Marcel. [1934] 1973. "Techniques of the Body." *Economy and Society* 2(1): 70–88.

McDonald, Hamish. 1980. *Suharto's Indonesia*. Blackburn, Australia: Fontana Books.

Messick, Brinkley. 1993. *The Calligraphic State: Textual Domination and History in a Muslim Society.* Berkeley: University of California Press.

Metcalf, Barbara. 1982. *Islamic Revival in British India: Deoband, 1860–1900.* Princeton: Princeton University Press.

Mitchell, Timothy. 1988. *Colonising Egypt.* Berkeley: University of California Press.

Nederburgh, I.A. 1888. "Vertaling eener 'Verzameling van inlandsche wetten.' " *Indisch Weekblad* (1292, 1293, 1294, 1296). Reprinted in 1919 in *Adatrechtbundels XVII* Series P No. 11 III: 152–176.

Needham, Rodney. 1958. "A Structural Analysis of Purum Society." *American Anthropologist* 60: 75–101.

van Niel, Robert. [1960] 1984. *The Emergence of the Modern Indonesian Elite.* Dordrecht: Foris Publications.

Niemann, G.K. 1889. "De Boegineezen en Makassaren: Linguistische en ethnologische studien." *BKI* 38: 74–88.

Nietzsche, Friederich. 1966. *The Genealogy of Morals.* In Walter Kaufman, ed. *Basic Writings of Nietzsche.* New York: Modern Library.

Niles, John. 1999. *Homo Narrans: The Poetics and Anthropology of Oral Literature.* Philadelphia: University of Pennsylvania Press.

Noer, Deliar. 1973. *The Modernist Muslim Movement in Indonesia, 1900–1942.* Singapore: Oxford University Press.

Noorduyn, J. 1956. "De Islamisering van Makasar." *BKI* 112: 17–266.

———. 1972. "Arung Singkang (1700–1765): How the Victory of Wadjo' Began." *Indonesia* 13: 61–68.

———. 1987a. "Makasar and the Islamization of Bima." *BKI* 143(2–3): 312–342.

———. 1987b. *Bima en Sumbawa.* VKIT 129. Dordrecht: Foris Publications.

Nurdin Daeng Magassing. [1933] 1981. *Riwayat Syekh Yusuf dan Kisah I Makkutaknang dengang Mannuntungi.* [Transliterated and translated by Djirong Basang into Bahasa Indonesia.] Jakarta: Departemen Pendidkan dan Kebudayaan.

Nuruddin ar-Raniri. [1642] 1983. *Khabar Akhirat Dalam Hal Kiamat.* Jakarta: Proyek Penerbitan Buku Sastra Indonesia dan Daerah.

Ochs, Eleanor and Lisa Capps. 2001. *Living Narrative: Creative Lives in Everyday Storytelling.* Cambridge: Harvard University Press.

O'Fahey, R.S. and Berndt Radtke. 1993. "Neo-Sufism Reconsidered." *Der Islam* 70: 52–87.

Ortner, Sherry. 1984. "Theory in Anthropology since the Sixties." *Comparative Studies in Society and History* 26(1): 126–166.

Pelras, Christian. 1979. "L'Oral et l'Écrit dans la Tradition Bugis." *Asie du Sud-Est et Monde Insulindien* 10: 271–297.

———. 1985. "Religion, Tradition and the Dynamics of Islamization in South Sulawesi." *Archipel* 29: 107–136.

———. 1996. *The Bugis.* Oxford: Blackwell Publishers.

Pelras, Christian. 1997. "La Première Description de Célèbes-sud en Francais et le Destinée Remarquable de Deux Jeunes Princes Makassar dans la France de Louis XIV." *Archipel* 54: 63–80.

———. 1998. "La conspiration des Makassar à Ayuthia en 1686." *Archipel* 56: 163–198.

Pemberton, John. 1994. *On the Subject of "Java."* Ithaca: Cornell University Press.

Qutb, Syed Shaheed. 1981. *Milestones.* Trans. S. Badrul Hasan. Karachi: International Islamic Publishers.

Rafael, Vicente. 1988. *Contracting Colonialism: Translation and Christian Conversion in Tagalog Society under Early Spanish Rule.* Ithaca: Cornell University Press.

Raffles, Thomas. 1830. *The History of Java.* London: John Murray.

Rahman, Fazlur. 1979. *Islam.* Chicago: Chicago University Press.

Ras, J.J. 1973. "The Pandji Romance and W.H. Rassers' Analysis of its Theme." *BKI* 129: 412–456.

Rassers, W.H. 1922. *De Pandji Roman.* Leiden.

Regeerings Almanak. *Regeerings-Almanak voor Nederlandsch-Indië.* Batavia: Landsdrukkerij.

Reid, Anthony. 1975. "Trade and the Problem of Royal Power in Aceh." In Anthony Reid and Lance Castles, eds. *Pre-Colonial State Systems in Southeast Asia.* Monographs of the Malaysian Branch of the Royal Asiatic Society No. 6. Kuala Lumpur: Perchetakan Mas Sdn. Bhd.

———. 1993. *Southeast Asia in the Age of Commerce.* Volume Two: *Expansion and Crisis.* New Haven: Yale University Press.

Richards, John. 1993. *The Mughal Empire.* Cambridge: Cambridge University Press.

Ricklefs, Merle. 1981. *A History of Modern Indonesia.* Bloomington: Indiana University Press.

Rizvi, S.A.A. 1965. *Muslim Revivalist Movements in Northern India in the Sixteenth and Seventeenth Century.* Agra University, Agra.

———. 1975. Religious and Intellectual History of the Muslims in Akbar's Reign. New Delhi: Munshiram Manoharlal Publishers.

———. 1978. *A History of Sufism in India, Volume I.* New Delhi: Munshiram Manoharlal Publishers.

———. 1983. *A History of Sufism in India, Volume II.* New Delhi: Munshiram Manoharlal Publishers.

Robinson, Francis. 1993. "Technology and Religious Change: Islam and the Impact of Print." *Modern Asian Studies* 27(1): 229–251.

Roeder, O.G. [1969] 1970. *The Smiling General: President Soeharto of Indonesia.* Second Revised Edition. Djakarta: Gunung Agung Ltd.

Rookmaker, H.R. 1924. "Oude en Nieuwe Toestanden in het Voormalige Vorstendom Bone." *De Indische Gids* 46(1): 397–417, 508–527. Amsterdam: J.H. Bussy.

Sahlins, Marshall. 1981. *Historical Metaphors and Mythical Realities: Structure in the Early History of the Sandwich Islands Kingdom.* Ann Arbor: University of Michigan Press.

———. 1985. *Islands of History*. Chicago: University of Chicago Press.

———. 1995. *How "Natives" Think: About Captain Cook, For Example*. Chicago: University of Chicago Press.

Salmon, David, ed. 1932. *The Practical Parts of Lancaster's Improvements and Bell's Experiment*. Cambridge: Cambridge University Press.

Schiller, A.A. 1955. *The Formation of Federal Indonesia 1945–1949*. The Hague: W. van Hoeve Ltd.

Schulte Nordholt, N.G. 1980. "The Indonesian Elections: A National Ritual." In R. Schefold, J.W. Schoorl, and J. Tennekes, eds. *Man, Meaning and History: Essays in Honor of H.G. Schulte Nordholt*. VKIT 89. The Hague: Martinus Nijhoff.

Scott, James. 1985. *Weapons of the Weak: Everyday Forms of Peasant Resistance*. New Haven: Yale University Press.

Siradjuddin Bantang. 1982. *Sinrilik Kappalak Tallung Batuwa*. Jakarta: Departemen Pendidikan dan Kebudayaan.

Skinner, C. 1963. *Sja'ir Perang Mengkasar by Entji' Amin*. The Hague: Martinus Nijhoff.

Snouck Hurgronje, Christian. 1906. *The Acehnese*. Two volumes. Leiden: E.J. Brill.

Southey, R. and C. Southey. 1844. *Life of the Rev. Andrew Bell*. London.

Sperber, Dan. 1975. *Rethinking Symbolism*. Cambridge: Cambridge University Press.

———. 1985. "Anthropology and Psychology: Towards an Epidemiology of Representations." *Man* (N.S.) 20: 73–89.

Starrett, Gregory. 1998. *Putting Islam to Work: Education, Politics and Religious Transformation in Egypt*. Berkeley: University of California Press.

Steenbrink, Karel. 1991. "Hamka en de Nadere Islamisering van Makassar." In Harry Poeze and Pim Schoorl eds. *Excursies in Celebes: Een Bundel Bijdragen bij het Afscheid van J. Noorduyn als Directeur-Secretaris van het KITLV*. VKIT 147. Leiden: KITLV Press.

von Stubenvoll, John. 1817. *History of the Island of Celebes*. Four volumes. Calcutta: Calcutta Gazette Press.

Sutherland, Heather. 1983. "Slavery and the Slave Trade in South Sulawesi, 1660s–1800s." In Anthony Reid, ed. *Slavery, Bondage and Dependency in Southeast Asia*. New York: St. Martin's Press, 263–285.

Tambiah, S.J. 1976. *World Conqueror and World Renouncer: A Study of Buddhism and Polity in Thailand against a Historical Background*. Cambridge: Cambridge University Press.

Taylor, Jean. 1983. *The Social World of Batavia: European and Eurasian in Dutch Asia*. Madison: University of Wisconsin Press.

Tideman, J. 1908. "De Batara Gowa op Zuid-Celebes." *BKI* 61: 350–390.

Toumey, Christopher. 1994. *God's Own Scientists: Creationists in a Secular World*. New Brunswick: Rutgers University Press.

Trimingham, J.S. 1971. *The Sufi Orders in Islam*. Oxford: Clarendon Press.

Tudjimah. 1987. *Syekh Yusuf Makasar*. Jakarta: Departemen Pendidkan dan Kebudayaan.

Turner, Victor. 1969. *The Ritual Process: Structure and Anti-Structure.* Harmondsworth: Penguin Books.

Usop, K. 1985. "Pasang ri Kajang: Kajian Sistem Nilai Masyarakat Amma Toa." In Mukhlis and Katherine Robinson, ed. *Agama dan Realitas Social.* Ujung Pandang: Lembaga Penerbitan Universitas Hasanuddin, 91–183.

van der Veer, Peter. 1999. "The Moral State: Religion, Nation and Empire in Victorian Britain and British India." In Peter van der Veer and Hartmut Lehmann, eds. *Nation and Religion: Perspectives on Europe and Asia.* Princeton: Princeton University Press, 15–43.

———. 2001. *Imperial Encounters: Religion and Modernity in India and Britain.* Princeton: Princeton University Press.

Voll, John. 1987. "Linking Groups in the Networks of Eighteenth Century Revivalist Scholars: The Mizjaji family in Yemen." In Nehemiah Levtzion and John Voll, eds. *Eighteenth-Century Renewal and Reform in Islam.* Syracuse: Syracuse University Press, 69–92.

———. 1988. "Scholarly Interrelations Between South Asia and the Middle East in the 18th Century." In P. Gaeffke and D. Utz, eds. *The Countries of South Asia.* Proceedings of the South Asia Seminar III (1982–1983). Philadelphia: Department of South Asia Regional Studies, 49–59.

van Vollenhoven, Cornelis. 1931. *Het Adatrecht van Nederlandsch-Indië, Volume II.* Leiden: E.J. Brill.

———. 1981. *Van Vollenhoven on Indonesian Adat Law.* KITLV Translation Series 20. The Hague: Martinus Nijhoff.

Voorhoeve, P. 1980. *Handlist of Arabic Manuscripts.* 2nd Edition. Codisces Manuscripti VII, Bibliotheca Universitatis Leidensis. The Hague: Leiden University Press.

Vredenbregt, J. 1962. "The Haddj." *BKI* 118: 91–154.

Warren, James. 1981. *The Sulu Zone, 1768–1898: The Dynamics of External Trade, Slavery and Ethnicity in the Transformationn of a Southeast Asia Maritime State.* Singapore: Singapore University Press.

Weber, Max. 1963. *The Sociology of Religion.* Boston: Beacon Press.

———. 1975. *Roscher and Knies: The Logical Problems of Historical Economics.* London: The Free Press.

———. 1978. *Economy and Society.* Berkeley: University of California Press.

———. 1985. *The Protestant Ethic and the Spirit of Capitalism.* London: Unwin.

Wittfogel, Karl. 1957. *Oriental Despotism.* New Haven: Yale University Press.

Wolhoff, G.J. and Abdurrahim. [1960]. *Sedjarah Gowa.* Jajasan Kebudajaan Sulawesi Selatan dan Tenggara.

Woods, John. 1999. *The Aqquyunlu: Clan, Confederation, Empire.* Revised Edition. Salt Lake City: University of Utah Press.

Woodward, Mark. 1989. *Islam in Java: Normative Piety and Mysticism in the Sultanate of Yogyakarta.* Tucson: University of Arizona Press.

van Wouden, F.A.E. [1935] 1968. *Types of Social Structure in Eastern Indonesia.* Trans. from the Dutch by R. Needham. The Hague: Martinus Nijhoff.

Wurtzburg, C.E. 1954. *Raffles of the Eastern Isles.* London: Hodder and Stoughton.

Zollinger, H. 1850. "Verslag van Eene Reis naar Bima en Soembawa." *Verhandelingen van het Bataviaasch Genootschap voor Taal-, Land- en Volkenkunde* 23–1. Batavia: Lange.

Zwemmer, S.M. 1925. "Islam at Cape Town." *The Moslem World* 15(4): 327–333.

Index

THOMAS GIBSON received his Ph.D. in Social Anthropology from the London School of Economics and is currently Professor and Chair of Anthropology at the University of Rochester. He began research in Island Southeast Asia in 1979 with fieldwork that resulted in a monograph called *Sacrifice and Sharing in the Philippine Highlands: Religion and Society among the Buid of Mindoro*. Since 1988, he has been conducting anthropological, historical, and literary research on symbolic knowledge and political authority in South Sulawesi, Indonesia. *Islamic Narrative and Authority in Southeast Asia* is the second in a series of monographs on this topic. The first of appeared under the title *And the Sun Pursued the Moon: Symbolic Knowledge and Traditional Authority Among the Makassar*. A final volume will examine the generation of conflicting experiences of the self and the world in Austronesian life-cycle rituals, Islamic mystical practices, and bureaucratic disciplines. Gibson's long-term goal is to develop a general theory of how diverse forms of symbolic knowledge and political authority interact in complex societies.

Printed in the United States
By Bookmasters